India Studies in Business and Economics

The Indian economy is considered to be one of the fastest growing economies of the world with India amongst the most important G-20 economies. Ever since the Indian economy made its presence felt on the global platform, the research community is now even more interested in studying and analyzing what India has to offer. This series aims to bring forth the latest studies and research about India from the areas of economics, business, and management science. The titles featured in this series will present rigorous empirical research, often accompanied by policy recommendations, evoke and evaluate various aspects of the economy and the business and management landscape in India, with a special focus on India's relationship with the world in terms of business and trade.

More information about this series at http://www.springer.com/series/11234

Madhusudan Ghosh • Debashis Sarkar
Bidhan Chandra Roy
Editors

Diversification of Agriculture in Eastern India

Springer

Editors
Madhusudan Ghosh
Department of Economics & Politics
Visva-Bharati
Santiniketan (Distt Birbhum)
West Bengal
India

Bidhan Chandra Roy
Dept of Agri Extension, Agri Economics & Agri Statistics
Institute of Agriculture, Visva-Bharati
Sriniketan (Distt Birbhum)
West Bengal
India

Debashis Sarkar
Dept of Agri Extension, Agri Economics & Agri Statistics
Institute of Agriculture, Visva-Bharati
Sriniketan (Distt Birbhum)
West Bengal
India

ISSN 2198-0012 ISSN 2198-0020 (electronic)
ISBN 978-81-322-1996-5 ISBN 978-81-322-1997-2 (eBook)
DOI 10.1007/978-81-322-1997-2
Springer New Delhi Heidelberg New York Dordrecht London

Library of Congress Control Number: 2014948550

Printed on acid-free paper

Springer is part of Springer Science+Business Media (www.springer.com)

Preface

Indian agriculture has been diversifying from cereals to high-value crops and livestock products in accordance with the changing consumption pattern in favour of livestock, fruits and vegetables. The production strategy has been changing to encourage diversification of the production system without sacrificing the basic obligation of ensuring food security. Diversification of agriculture generates greater employment opportunities and higher incomes for farm households. Indian agriculture in general and Eastern India in particular, have been facing the challenges of the new economic regime, besides the usual problems of rising population, unemployment and poverty; declining investment in agriculture and degradation of natural resources. Diversification of agriculture may help to overcome these overriding problems in a more competitive environment.

This edited volume examines various aspects of agricultural diversification in Eastern India, namely, the rationale and extent of diversification, the nature and problems of diversification, food and livelihood security through diversification, etc. These issues are discussed against the background of significant structural transformation of the Indian economy from agriculture to non-agriculture (services and industry) and changing cropping pattern from cereals to non-cereals associated with changing consumption pattern.

This volume includes 19 chapters, besides an Introduction, covering various aspects of diversification of agriculture in Eastern India with special reference to the states of Assam, Bihar, Jharkhand, Orissa and West Bengal. The papers are organised into three parts, each part including a set of articles dealing with a particular issue of agricultural diversification. Earlier versions of the papers were presented at the 'Regional Seminar on Diversification of Agriculture in Eastern India', organized by the Agro-Economic Research Centre, Visva-Bharati in collaboration with the Indian Society of Agricultural Economics (ISAE), Mumbai and National Bank for Agriculture and Rural Development (NABARD) during March 23-25, 2012. We are thankful to ISAE and NABARD for sponsoring the Seminar.

We are also thankful to B. S. Bhandari, Kazi MB Rahim and Sibaji Chakraborti for their help and suggestions, and to the staff members of the Agro-Economic Research Centre, Visva-Bharati for their assistance in organizing the Seminar.

Visva-Bharati, Santiniketan

Madhusudan Ghosh
Debashis Sarkar
Bidhan Chandra Roy

Contents

Contributors

H. N. Atibudhi Department of Agricultural Economics, Orissa University of Agriculture & Technology, Bhubaneswar, Orissa, India

Debisree Banerjee Institute of Development Studies Kolkata (IDSK), Kolkata, West Bengal, India

G. D. Banerjee Vivekanand Education Society's Institute of Management Studies and Research, Mumbai, Maharashtra, India

Sarda Banerjee Mumbai, Maharashtra

B.K. Bera Department of Agricultural Economics, Bidhan Chandra Krishi Viswavidyalaya, Mohanpur, Nadia, West Bengal, India

Uttam Bhattacharya Institute of Development Studies Kolkata (IDSK), Kolkata, West Bengal, India

Uttam Kumar Bhattacharya Institute of Development Studies Kolkata (IDSK), Kolkata, West Bengal, India

M. K. Bhowmick Rice Research Station (Government of West Bengal), Hooghly, West Bengal, India

P. K. Biswas Institute of Agriculture (Palli Siksha Bhavana), Visva-Bharati, Birbhum, West Bengal, India

Tapas Kumar Chakrabarty 'Purbachal', Andrews Pally, Birbhum, West Bengal, India

Formerly, in Reserve Bank of India, Mumbai, India

A. Jha Chakraborty Department of Agricultural Economics, Bidhan Chandra Krishi Viswavidyalaya, Mohanpur, Nadia, West Bengal, India

C. L. Dadhich Indian Society of Agricultural Economics, Mumbai, Maharashtra, India

B. Das Agricultural Sector Officer, Kadamtala, Agril. Sub-Division, North Tripura, Govt. of Tripura, Tripura, India

K. K. Das Department of Agricultural Economics, Uttar Banga Krishi Viswavidyalaya, Pundibari, Cooch Behar, West Bengal, India

Sima Das Department of Agricultural Economics, Assam Agricultural University, Jorhat, Assam, India

K.K. Datta National Dairy Research Institute (Deemed University), Karnal, Haryana, India

B. Duary Institute of Agriculture (Palli Siksha Bhavana), Visva-Bharati, Birbhum, West Bengal, India

Madhusudan Ghosh Department of Economics and Politics, Visva-Bharati, Santiniketan Birbhum, West Bengal, India

Basant Kumar Jha Agro-Economic Research Centre For Bihar and Jharkhand, T.M. Bhagalpur University, Bhagalpur, Bihar, India

K.G. Karmakar S.P. Jain Institute of Management & Research, Mumbai, Maharashtra, India

Dilruba Khatun Department of Economics, Kulti College, Kulti, West Bengal, India

S. Maji Department of Agricultural Economics, Bidhan Chandra Krishi Viswavidyalaya, Mohanpur, Nadia, West Bengal, India

Subhasis Mandal Central Soil Salinity Research Institute, Canning Station, West Bengal, India

G. Mula NAIP, Component-3, Department of Agricultural Economics, Uttar Banga Krishi Viswavidyalaya, Cooch Behar, West Bengal, India

A.K. Nandi Department of Agricultural Economics, Bidhan Chandra Krishi Viswavidyalaya, Mohanpur, Nadia, West Bengal, India

R. K. Panda Nabakrushna Coudhury Centre for Development Studies, Bhubaneswar, Odisha, India

B. Paramanik NAIP, Component-3, Department Soil Science and Agril. Chemistry, Uttar Banga Krishi Viswavidyalaya, Pundibari, Cooch Behar, West Bengal, India

P. S. Patra NAIP, Component-3, Department of Agronomy, Uttar Banga Krishi Viswavidyalaya, Cooch Behar, West Bengal, India

D. Rai Department of Agricultural Economics, Uttar Banga Krishi Viswavidyalaya, Pundibari, Cooch Behar, West Bengal, India

Bidhan Chandra Roy Department of Agricultural Extension, Agricultural Economics and Agricultural Statistics (EES), Palli Siksha Bhavana (Institute of Agriculture), Visva-Bharati, Sriniketan, West Bengal, India

Debajit Roy AERC, Visva-Bharati, Santiniketan, West Bengal, India

T. N. Roy Department of Agricultural Economics, Uttar Banga Krishi Viswavidyalaya, Pundibari, Cooch Behar, West Bengal, India

B.B. Sahoo National Bank for Agriculture and Rural Development (NABARD), Bangalore, Karnataka, India

Debashis Sarkar Department of Agricultural Extension, Agricultural Economics and Agricultural Statistics (EES), Palli Siksha Bhavana (Institute of Agriculture), Visva-Bharati, Sriniketan, West Bengal, India

S. C. Sarker Department of Agricultural Economics, Uttar Banga Krishi Viswavidyalaya, Cooch Behar, West Bengal, India

Shiv Raj Singh Shri G.N. Patel Dairy Science & Food Technology College, SDAU, S.K. Nagar, Gujarat, India

Rajiv Kumar Sinha Agro-Economic Research Centre for Bihar and Jharkhand, Old P.G. Campus, T.M. Bhagalpur University, Bhagalpur, Bihar, India

K.C. Talukdar Department of Agricultural Economics, Assam Agricultural University, Jorhat, Assam, India

Udeshna Talukdar Department of Economics and Sociology, Punjab Agricultural University, Ludhiana, Punjab, India

About the Editors

Madhusudan Ghosh is a Professor of Economics at the Department of Economics & Politics, Visva-Bharati, Santiniketan, West Bengal (India). He was a Ford Foundation Post-doctoral Fellow at the University of California, Berkeley (USA), and a Visiting Professor at Soka University, Tokyo (Japan). He was awarded Post-doctoral Fellowship in Economics by the Ford Foundation, New Delhi, and Dr. D.K Desai Memorial Prize (four times) by the Indian Society of Agricultural Economics, Mumbai. He is an active researcher and has been teaching for the last 34 years. He has published books such as *Liberalization, Growth and Regional Disparities in India* (Springer, 2013), *Economic Reforms and Indian Economic Development: Selected Essays* (Bookwell, 2008) and *Rural Development in India: Challenges and Prospects* (Co-edited; Serials Publications, 2013), and numerous research papers in reputed national and international peer-reviewed professional journals. He has also worked on research projects sponsored by University Grants Commission, New Delhi.

Debashis Sarkar is an Associate Professor of Agricultural Economics at the Department of Agricultural Extension, Agricultural Economics & Agricultural Statistics, Institute of Agriculture, Visva-Bharati, Sriniketan. He also served as Director, Agro-Economic Research Centre, Visva-Bharati for two years. He has completed many research projects sponsored by University Grants Commission (UGC), National Centre for Agricultural Economics and Policy Research (NCAP), Directorate of Jute Development, Government of India, Directorate of Crop Production, Government of India, and Directorate of PPQ&S, Government of India. He has published seven books (four edited) and many articles in reputed national and international peer-reviewed journals.

Bidhan Chandra Roy is an Associate Professor of Agricultural Economics at the Department of Agricultural Extension, Agricultural Economics & Agricultural Statistics, Institute of Agriculture, Visva-Bharati, Sriniketan. Before joining Visva-Bharati, he served as a Scientist at National Centre for Agricultural Economics and Policy Research (NCAP), New Delhi. He has developed an interactive policy dialogue model to assess the water and food security scenarios at agro-ecological region level in India. He has published a book and many articles in reputed national and international peer-reviewed journals.

About the Editors

List of Figures

List of Figures

List of Tables

Chapter 1
Introduction

Madhusudan Ghosh, Debashis Sarkar and Bidhan Chandra Roy

Diversification of agriculture is considered as an important strategy to overcome the challenges faced by many developing countries. Diversification of agriculture means developing a larger-number crop mix or enterprise mix in favour of high-value and more remunerative enterprises. It may be of different forms such as supplementing farm incomes with non-farm incomes, increasing the number of crops grown and types of livestock reared, or use of resources in diverse farm enterprises. The prominent arguments in favour of diversification of agriculture are to increase farm income, generate additional employment, stabilize farm income overtime, and to conserve natural resources.

Diversification has been pursued in many countries as a way to improve the long-term viability of agriculture by enhancing the profitability and overall stability of the sector. The shift to other crops or economic activities, however, has not been an easy undertaking, particularly for small farmers. Government assistance in terms of more supportive policies and better infrastructure has played a significant role in the promotion of diversification programmes. With globalisation further stimulating trade, diversification of agriculture afforded greater opportunities for expanding the range of agricultural products that one can market abroad. However, expanded trade has also brought with it higher competition and hence the need to focus diversification programmes on agricultural activities where they have a competitive advantage.

Indian agriculture has been diversifying from cereals to high-value crops and livestock products since the early 1990s. With high economic and population

M. Ghosh (✉)
Department of Economics and Politics, Visva-Bharati, Santiniketan, Birbhum, West Bengal 731235, India
e-mail: msghosh123@rediffmail.com

D. Sarkar · B. C. Roy
Department of Agricultural Extension, Agricultural Economics and Agricultural Statistics (EES), Palli Siksha Bhavana (Institute of Agriculture), Visva-Bharati, Sriniketan, West Bengal, India
e-mail: debashis.sarkar@visva-bharati.ac.in

B. C. Roy
e-mail: bcroy1969@rediffmail.com

M. Ghosh et al. (eds.), *Diversification of Agriculture in Eastern India*,
India Studies in Business and Economics, DOI 10.1007/978-81-322-1997-2_1

growth, the dietary pattern in India is also changing fast. The food basket is diversifying in favour of livestock, fruits, and vegetables. In response to these demands, the crop mix is changing in favour of commercial crops and from low- to high-elasticity commodities. Yet, producing additional diversified food is a major challenge when resources are limited and degrading. To meet the challenge, the production strategy should be to encourage diversification of the production system without sacrificing the basic obligation of ensuring food security.

The changing cropping pattern is thought to be due to the interactive effects of many factors, such as: (1) resource-related factors like irrigation, rainfall, and soil fertility; (2) technology-related factors including not only seed, fertilizer, and water technologies but also those related to marketing, storage, and processing; (3) household-related factors including food and fodder self-sufficiency requirement, and investment capacity; (4) price-related factors covering output and input prices as well as trade policies and other economic policies that affect these prices either directly or indirectly; (5) institutional and infrastructure-related factors covering farm size and tenancy arrangements, research, extension and marketing systems, and government regulatory policies.

The relative importance of these factors has, however, been changing over time. Indian agriculture is increasingly being influenced by economic factors. This is not surprising because irrigation expansion, infrastructure development, penetration of rural markets, development and spread of short duration and drought resistant crop technologies have contributed to minimizing the role of non-economic factors in crop choice of even small farmers. Moreover, the reform initiatives undertaken in the context of ongoing agricultural liberalisation and globalisation policies since the early 1990s have strengthened the role of price-related economic incentives in determining crop composition both at the micro and macro levels. Obviously, such a changing economic environment will also ensure that government price and trade policies will become powerful instruments for directing area allocation decisions of farmers, aligning thereby the crop pattern changes in line with the changing demand-supply conditions. In a condition, where agricultural growth results more from productivity improvement than from area expansion, the increasing role that price-related economic incentives play in crop choice can also pave the way for the next stage of agricultural evolution, where growth originates more and more from value-added production.

It has been reported that agricultural diversification generates higher incomes for farm households, and greater employment opportunities, particularly for women. The shift in land area from cereals to vegetables, in particular, has enhanced employment opportunities in rural areas. (Joshi 2005; Joshi et al. 2004, 2006; Vyas 1996). However, the combination of a large number of small farmers, poor rural infrastructure, and fragmented and underdeveloped markets complicates establishment of efficient and equitable links between farmers and the diverse, emerging domestic market. Indian agriculture, in general and eastern India in particular, is facing the complex challenge of the new economic regime, besides the usual problems of rising population, unemployment and poverty, declining investment in the agricultural sector, and degradation of natural resources. Diversification of agriculture may help to overcome these overriding problems in a more competitive environment.

Diversification of agriculture also assumes significance in the context of significant structural change of the Indian economy from agriculture to non-agriculture (services and industry); the contribution of the agricultural and allied sector to gross domestic product (GDP) declined from 32.2% in 1990–1991 to about 15% in 2009–2010. However, despite the declining share of agriculture in GDP, the importance of this sector can hardly be over emphasised in view of the fact that this sector still absorbs an overwhelming majority of the workforce. Naturally, the standard of living of a large section of the rural population depends significantly on the performance of this sector.

Agricultural diversification assumes special significance in eastern India because of the fact that the planning commission has given special emphasis in bringing second green revolution to this region. Diversification of agricultural production and marketing offers the opportunity to strengthen lagging growth in farm output and rural employment in this part of India. However, achieving diversified growth with equity requires new measures to increase investment and provide the market institutions needed to develop eastern India's inefficient food processing and marketing sectors, and to ensure that the transformation to high-value agriculture is inclusive of the region's large number of marginal and small farmers as well as landless labourers.

In this context, it is useful to examine the following emerging issues in agricultural diversification in eastern India: (1) rationale for diversification; (2) nature and extent of diversification in various production systems; (3) scope for horizontal and vertical diversification towards high-value crops as well as non-farm activities; (4) determinants of diversification; (5) constraints (agro-ecological/technological/socioeconomic) in realizing the potential benefits of diversification; (6) possible impacts of diversification on different livelihood groups; (7) institutional arrangements (e.g., contract farming, micro-finance, land acquisition norms, and crop insurance) to promote diversification; and (8) policy interventions in terms of price protection, development of rural infrastructure particularly market and storage, land ceiling and tenancy act needed to promote diversification of agriculture.

The chapters included in this edited volume examine various aspects of agricultural diversification in eastern India with particular reference to the states of Assam, Bihar, Jharkhand, Orissa, and West Bengal. This volume is organised in three parts, each part dealing with a set of emerging issues in agricultural diversification. Earlier versions of the volume were presented at the 'Regional Seminar on Diversification of Agriculture in Eastern India', organised by the Agro-Economic Research Centre, Visva-Bharati, in collaboration with the Indian Society of Agricultural Economics (ISAE), Mumbai, and National Bank for Agriculture and Rural Development (NABARD) during March 23–25, 2012. The major findings of the volume are summarised in the following parts.

1.1 Part I

Part I includes seven chapters on various issues relating to rationale and extent of agricultural diversification. K. G. Karmakar and B. B. Sahoo, in their chapter on 'Green Revolution in Eastern India', have discussed the role of various stakeholders in the overall agricultural growth strategy in the eastern region. They argue that the eastern India is relatively backward in terms of crop diversification and crop productivity. Non-viability of the small farms, inadequacies in the available inputs, extension services, marketing systems, and infrastructure have been found to be the primary reasons behind the continuing low productivity and poverty syndrome and enhanced regional disparities in this region. Changes in institutional arrangements in the land tenure system such as land leasing, contract farming, and corporate farming could make small farms economically viable. Further, by involving the corporate sector in agri-business and value chain innovations, meaningful employment opportunities could be generated for educated youth. With the State Governments as facilitators in eastern India, the small and marginal farmers, the corporate sector and all financial institutions can work together towards mutual prosperity, besides ensuring food security. By forging strategic alliances with key stakeholders, the corporate sector can provide a sound business framework, which could be a panacea for the ills of the agriculture sector in general and the small and marginal farmers in eastern India, in particular. Recognising the gap between available potentials and actual growth achieved in agriculture in the eastern region, the action plan suggested here to integrate small and marginal farmers, quality inputs supply, output markets, and supply chains could be useful in the utilisation of fund created for bringing Green Revolution in eastern India. The chapter on 'Diversified Sustainable Agriculture in Eastern India' by T. K. Chakrabarty has stressed that farmers' mind set needs to be changed towards the knowledge-based thinking like that of farmers of agro rich states of the country. The problems of marginal and small-scale farmers of the eastern region can be addressed through the state initiatives supporting contract farming. Diversification of agriculture in eastern zone was towards non-cereal products and to some extent towards livestock. Presently, demand for protein-based products like meat, egg, milk, and fish as well as fruits and vegetables has increased substantially with rise in income, assisted by demographic dividend and policy initiatives. This has offered huge opportunity to the unexplored fertile agricultural area of eastern zone to sustain diversified agricultural activities. The need of the hour is the young farmers' mind set towards application of modern agro facilities like contract farming, trading in futures and other facilities from finance and insurance through collaborative efforts.

Discussing various dimensions of crop diversification in India, G. D. Banerjee and S. Banerjee ('Crop Diversification: An Exploratory Analysis') argue that crop diversification (horizontal and vertical) is one of the best options to increase farm income leading to food, nutrition and ecological security as well as poverty alleviation. Therefore, greater attention should be paid to crop diversification by the government. Several steps can be taken to reduce risks and improve marketing facilities through improved roads and communications, construction of wholesale markets, etc. Access by farmers, private traders, and exporters to credit also needs to be improved. Efforts should be made to identify high specialty crops, new crops, off-season varieties, and production systems to open up new opportunities for farmers. The promotion of multipurpose species would also be useful for diversification of agro-processing on small-scale at local and national level for productivity enhancement and expanded employment opportunities. Again, there is a need for improved seed and other planting materials for effective crop diversification. One major concern is the high post-harvest losses of crop produce, particularly in horticultural crops. The government should take initiatives to minimize such losses. The private sector can play a major role in the development of modern agro-enterprises to infuse capital and technology into diversified cropping systems for effective commercialisation for long-term sustainability. Since crop diversification is an important element of poverty alleviation, income generation, equity, and natural resource conservation, a well-designed mechanism has to be developed through the participation of the local governments. There is also a need for development of an information database on crop diversification for policy makers, farmers, consumers, and other stakeholders.

Based on secondary data, the study on 'Trend and Pattern of Crop Diversification in Odisha' by R.K. Panda examines the trend and pattern of crop diversification at the state and regional levels. In view of the predominance of small landholdings in the state, the study also examines the association of small holders in the cultivation of high-value crops (HVCs). It also outlines the recent policy initiatives taken up by the state government towards crop diversification. The findings of the study reveal a negative growth rate in the area under the cultivation of paddy, fibres, and tobacco, and a positive growth rate in acreage under pulses, oilseeds, spices, vegetables, sugarcane, and fruits at the state level from 2001/2002 to 2009/2010. Particularly, the trend growth rate in area under sugarcane and pulses was quite impressive during the period. At the regional level, the agro-climatically better-off region (coastal plain) does not show better performance in adaptation to commercial and high-value crops. The growth in acreages under oilseeds, fibres, fruits, vegetables, and sugarcane in this region lags behind other regions. The concentration of small holders is found to be higher in coastal plain as compared to other regions. The participation of small holders with varying degree across regions in crop diversification particularly with regard to fruits and vegetables is noteworthy. The recent policy initiatives taken by state government in raising crop diversity in the state is also noteworthy. However, these policies need to be implemented effectively by improving rural infrastructures and raising institutional support to small farmers who are found playing positive role in diversification process. H. N. Atibudhi ('Pattern of Agricultural Diversification in Odisha') analyses the trends and patterns of agricultural diversification in Odisha. He observes that there has been a significant change in the cropping pattern in the past few decades in the country as a whole as well in Odisha. The share of cereals in gross cropped area (GCA) was highest amongst other crops from 1970/1971 to 2007/2008. Moreover, the area devoted to food grains (cereals and pulses) was much higher in both the state and all India levels. However, diversification away from food grains was more prominent in Odisha in comparison to all India. In the state, diversification was found to be most remarkable towards pulses and oilseeds, though after 2000, the area under oilseeds has declined. There have been remarkable changes in the relative shares of various crops (with significant contribution of fruits and vegetables) in the gross value of crop output from agriculture in the past few decades. The uncertainty in the crop production sector has warranted special policy interventions for strengthening the ailing livestock sector, which can provide supplementary incomes to the farmers and can contribute to the gross state domestic product significantly.

In the chapter on 'A Study on the Extent of Crop Diversification in West Bengal', S. Maji et al. observe that the agricultural sector in West Bengal has been gradually undergoing diversification in favour of high-valued food crops like potato, oilseeds, fruits, and vegetables. However, the pace of diversification has not been as fast as needed for speeding up of growth in agriculture. They argue that the pattern of diversification in the state is due to expansion effect. Based on primary data collected from *Katarni* paddy growing cultivators from Bhagalpur and Banka districts of Bihar, the chapter on 'A Study of Diversification of Katarni to HYV Paddy in Bihar' by B.K. Jha and R.K. Sinha analyses the trend and causes of diversification within crop (from '*Katarni* Paddy' to HYV paddy) in Bihar. They observe that even though *Katarni* paddy is unique and marvelous for its taste and flavour, uncomparable with any other paddy in the world, it has been facing the threat of extinction. Since 1991/1992, the area under *Katarni* paddy has started declining significantly mainly due to the constraints/reasons like (1) unchecked excavation of sand from the river *Chandan,* leading to declining water retention capacity of the river and escalation in irrigation cost; (2) erosion of genetic purity of *Katarni* paddy; (3) higher productivity of other varieties of HYV paddy; (4) lower productivity of *Katarni* paddy and in proportion to that, lower prices paid to the actual growers; and (5) declining domestic and global demand owing to selling of adulterated *Katarni* paddy by local traders/middlemen. Based on their field level experience, they have suggested some suitable measures to increase area under this aromatic rice.

1.2 Part II

The nature and problems of agricultural diversification are discussed in eight chapters included in Part II of this volume. In the chapter on 'Role of Dairying in Diversification of Indian Agriculture', C.L. Dadhich argues that diversification of agriculture is imperative to ensure among others nutritional security and smooth flow of rural income stream. While value of agriculture output is highly volatile, value of livestock output in general and value of milk output in particular is steady and stable. Dairy sector is one of the main drivers of growth of Indian agriculture. The importance of this sector cannot be overemphasised in the context of diversification of agriculture. Demand-led white revolution made rapid strides in majority of Indian states but largely by-passed eastern region. While most of the states registered shift in the composition of livestock in favour of dairy stock, the eastern states did not witness perceptible shift in its composition. The study brings to the fore that preponderance of low yielding indigenous cattle in dairy herd has adversely impacted the participation of this region in white revolution. This apart inadequate infrastructure has also caused untold damage to the growth of dairy sector and consequently non-diversification of agriculture. However, limited yet satisfactory induction of crossbreeding programme of cattle in the region indicates huge growth potential for dairy development. It does without saying that upgradation of indigenous cattle if not the exotic crossbreeding programme in a big way will place dairy sector on fast growth trajectory and go a long way in diversification on agriculture in the region.

In the chapter on 'Crop Diversification Through Oilseeds in Eastern India', M. K. Bhowmick et al. make a critical review on various dimensions of diversification in cropping systems through oilseeds in eastern India and analyse the advantage of crop diversification with oilseeds in marginal ecosystems to make the cropping enterprise as a profitable venture avoiding risks. They argue that Indian oilseeds sector, the fourth largest in the world after USA, China, and Brazil, is at crossroads with a wide gap between country's oilseeds production and escalating demand and mounting imports of vegetable oils causing a drain on the foreign exchange reserves. Hence, it is of immediate necessity to bridge the gap between demand and supply of oilseeds in the country, especially in the eastern states. Crop diversification through suitable oilseeds with appropriate management practices in areas, where cereal-dominating production system is in vogue, may be an effective option for mitigating or minimizing the present-day problems, besides making a sustainable improvement in oilseeds production. P. K. Biswas and M. K. Bhowmick ('Crop Diversification in North Eastern Plain Zone of India') have shown that the rice-based cropping systems have threatened agricultural sustainability by causing imbalance in soil nutrient and water availability, frequent outbreak of insect pests, disease epidemics, etc. Thus, in the fragile ecosystem and poor farm resource base, crop diversification on rice-based cropping systems is of utmost importance with a view to maintain a balance between crop intensification and sustainability of the production system. Declining factor productivity of cereal-based intensive cropping systems warrants diversification of cereals with pulses in India, particularly in the northeastern plain zone (NEPZ). Growing of *rabi* pulses in lowland rice-fallows of this zone holds a great promise to enhance pulse production without the risks of high input agriculture.

Based on household survey conducted in two districts of West Bengal, one representing a more diversified agriculture (Burdwan) and the other less diversified one (Purulia), the study on 'Crop Diversification in West Bengal: Nature and Constraints' by D. Khatun and B. C. Roy examines the nature and constraints to crop diversification among different livelihood groups. The findings of the study show that the level of crop diversification varies across the regions and different livelihood groups. On an average, while the households in Burdwan district grow more than three crops, with the highest number of crops being grown by the cultivator group followed by the salaried group, the households in Purulia district grow at most two crops, with as high as 41% of sample households growing only one crop per annum. The low level of crop diversification in Purulia was found to be due to distress induced by the agro-climatic factors, particularly due to erratic rainfall pattern and lack of any kind of irrigation facilities in the study area. The principal constraints faced by the rural households are of various kinds. While most of them are socio-economic in nature, some constraints are of agro-ecological nature, and there are few, which are technical or institutional in nature like non-availability of quality seeds resistant to extreme climates, and pests and diseases, etc. Spatial variation leads to cross sectional heterogeneity thereby influencing diversification pattern. Property rights in productive assets such as land and livestock, labour availability, and access to credit differs across livelihood groups. Therefore, though all the livelihood groups faces these constraints because of poor asset base, the severity of the constraints are more for the landless labourer groups and least for the resource-rich salaried class. High volatility in prices, absence of market, and lack of access to technical knowhow are the main constraints faced by the cultivators group. A large proportion of small and marginal farmers gain livelihoods through production on small pieces of land. For these households, timely availability or access to credit and improved methods of production are quite critical for their livelihood.

The chapter on 'Problems of Crop Diversification in West Bengal' by D. Banerjee and U. K. Bhattacharya argues that crop diversification helps to maximise the utilisation of scarce land resource, increase productivity, and reduce risk in agriculture. In West Bengal, there is a scope for increasing crop diversity with increased cultivation of *boro* rice, potato, different oilseeds, pulses, and horticultural crops. However, even though cropping diversity can reduce several limitations and risks involved in traditional methods of agriculture, it cannot eliminate the risk element completely. Agricultural productivity in West Bengal largely suffers due to small farm size. Changes in cropping pattern intensify the need for stronger financial institutions. Owing to the prevalence of small and marginal farmers in West Bengal, over-dependence on credit has become a problem for the state. In order to implement and sustain the benefits from crop diversification, technological and institutional supports are required. Agricultural insurance could be an important tool towards effective crop diversification programme. Agricultural insurance provides coverage to the farmers for any loss occurring in agriculture. The scheme extends various kinds of insurance apart from different yield-based crop insurance schemes that protects farmers in case of crop failure due to natural hazards, man-made hazards, perils and risks due to technological changes, and change in economic policies pursued by the Government. Agricultural insurance could also help the farmers to take risk involved with technological changes and change in cropping pattern protecting the probable financial risks. The study reveals that insurance in agriculture is less explored area and there is further scope in the state of West Bengal in its effort towards reducing the risks associated with crop diversification. It has stressed to explore the methods through which the benefits could reach to the needy ones. Based on a micro-survey in some agro-economic regions in West Bengal, the chapter on 'Factors Influencing the Extent of Diversification in West Bengal' by Debajit Roy, examines the conditions under which there has been greater diversification of cropping pattern, and the factors influencing the farmers' decision regarding diversification. The study finds that the areas, endowed with assured water supply at cheap rates (mostly canal irrigation), concentrate more on production of cereals and traditional crops showing lower extent of crop diversification. However, in the areas, where farmers have to depend on private sources of water at high cost due to non-availability of publicly supplied irrigation system, they diversify away from water-intensive cereals and other traditional crops towards various other high-value crops showing greater degree of diversification. Diversification in cropping pattern has occurred more in those cases, where farm households are in a position to provide more family labour for cultivation.

In their chapter on 'Policy Intervention in West Bengal Agriculture: Role of Diversification', S. R. Singh et al. argue that besides producing food grains eastern India has great potential to produce several high-value commodities like horticulture, livestock, and fisheries to accelerate the growth of agricultural output. However, one of the key impediments to fostering the agricultural growth in this region is the small and marginal production unit of the majority of the farmers. The small scale of production unit can produce these high-value commodities with high to moderate production efficiency, but poor marketing efficiency. Farming units are usually confronted with many unpredictable uncertainties ranging from climatic vagaries to market price fluctuations. The degree of uncertainty is greater for the small and marginal landholders, where the farmers do not have access to basic information on various risks including loss of assets and income. Keeping in view of the opportunities and prospects of agricultural growth, the chapter focuses on the ways and means of agricultural development in West Bengal, which may help to accelerate the rural income and household level food and nutritional security. Farming in West Bengal is individual-driven and unorganised, with the average size of holding being 0.82 ha, much lower than the national average of 1.33 ha. Therefore, individual farmers, with very small marketable surplus of produce, have to pay market price for all farm inputs and other basic utilities and consumable items. There is, therefore, a need to organise a vastly unorganised farming community in such a way as to help them to gain from the market economy. Smallholders are competitive in high-value agricultural activities because of the availability of family labour and their ability to compete in local markets. However, as production and marketing systems evolve, support to smallholders to provide efficient input services, links to output markets, and risk mitigation measures will be important, if they are to provide higher-value products. Innovative public support and links to the private sector will be required for the poor to adapt and benefit from the emerging systems.

In the chapter on 'Causes of Agricultural Diversification in Bihar and Jharkhand', R. K. Sinha argues that agricultural diversification in India has been towards a continuous increase in the share of allied activities and decline in the share of crop sector since the late sixties. Within crop sector, the trend of diversifications has, however, been changing periodically. In spite of these periodic shifts in crop acreage, proportionate area under fine cereals had been increasing over a long period. Though in some of the states, this trend continues, the recent trend in crop diversification shows decline in percent area under fine cereals and increase in percent area under non-foodgrain crops. Empirical evidence shows that crop diversification in Bihar has been towards pulses, fibre crops, oilseeds, cereals, and sugarcane during the period from 2000/2001 to 2008/2009. During the same period, farmers in Jharkhand have been diversifying towards oilseeds, pulses, maize, and wheat. Based on these findings he has suggested that in order to generate additional income by enhancing employment opportunities in rural areas on sustainable basis, the farmers need to be encouraged to undertake those non-food crops, which have greater potentiality of value addition.

1.3 Part III

Part III, consisting of four chapters, has discussed the issues regarding food and livelihood security through agricultural diversification. K. C. Talukdar et al., in their chapter on 'Diversification and Food Security in the North Eastern States of India', examine the impact of crop and agricultural farm diversification on food security in northeastern (NE) states of India. Using various indices of diversification to state level data for the period from 2001/2002 to 2011/2012, they evaluate crop and agricultural farm diversification across the states. In general, the magnitude of crop diversification was found to have declined across the states, but agricultural farm diversification has increased due to impressive growth in livestock, fisheries, and fruits and vegetables. It was observed that except fruits and vegetables, all other food grains were in deficit in the states measured in terms of their per capita availability per day. The gap between requirement and availability was higher in Assam and Sikkim. In terms of nutrients, protein was in surplus in all the states of NE India, while energy in terms of kilocalories was in deficit in the states due to low productivity of food grains. The study has indicated that the NE states need to increase productivity of food grains and oilseeds to reduce the food import bill of the region.

S. C. Sarker et al., in their chapter on 'Impact of Improved Agro-Techniques Towards Food and Livelihood Security', evaluate the impact of a sub-project on 'Sustainable Rural Livelihood Security' (SRLS) that has been in progress since 2008 in four districts of West Bengal (viz. Uttar Dinajpur, Dakshin Dinajpur, Malda, and Murshidabad) under National Agricultural Innovation Project (NAIP). The research process began with development of knowledge-management strategy and progressed further with technology intermediation and simultaneous ownership building activities. Interventions on new crop sequencing, resource conserving technologies and varietal replacement have resulted in 23.25 % increase in cropping intensity from baseline value of 142.20 %. As cumulative effect of integration of improved crop husbandry, new crop sequencing, and replacement of varieties, impressive increase in productivity has been achieved for all major cereals (49.63 %), oilseeds (43.02 %), pulses (58.44 %), and potato (23.85 %) to effectively address food insecurity issue of around 64.26 % local below poverty line (BPL) families.

In the chapter on 'Floriculture in West Bengal in Augmenting Income and Export', T. N. Roy et al. explore the possibilities of value addition of multi-utilities floricultural products in eastern India in general and West Bengal in particular. West Bengal has a unique and varied agro-climate condition, which facilitates to raise temperate to humid and sub-tropical flowers. It is an eco-friendly crop with high potential of income and employment generation for small and marginal producers. They observe that marketing of perishable crop like flowers always requires special arrangement. However, as the value chains of cut and loose flowers are mostly unorganised, efforts should be made to increase buyers' awareness and to improve infrastructural facilities. K. K. Das et al. ('Role of Muga Culture in Diversification Strategy') have argued for diversification of agriculture towards non-food grain and high-value commodities, because of its immense potential for income augmentation, employment generation, poverty alleviation, and export promotion. Muga silk, one of the three non-mulberry silks (*Muga, Eri, Tasar*), produced by the silkworm *Antherea assamensis,* is unique for its golden-yellow colour. This enterprise has experienced growing popularity because of its potential for providing high economic return and employment throughout the year.

References

Joshi PK (2005) Crop diversification in India: nature, pattern and drivers. National Centre for Agricultural Economics and Policy Research, New Delhi. http://www.adb.org/Documents/Reports/Consultant/TAR-IND-066/Agriculture/joshi.pdf. Accessed 15 March 2012

Joshi PK, Gulati A, Birthal PS, Tewari L (2004) Agricultural diversification in South Asia: patterns, determinants and policy implications. Econ Polit Wkly 39(24):2457–2468

Joshi PK, Birthal PS, Minot N (2006) Sources of agricultural growth in India: role of diversification towards high value crops, MTID Discussion Paper No. 98, IFPRI, Washington

Vyas VS (1996) Diversification in agriculture—concept, rationale and approach. Indian J Agric Econ 51(4):636–643

Part I
Rationale and Extent of Agricultural Diversification

Chapter 2
Green Revolution in Eastern India

K. G. Karmakar and B. B. Sahoo

2.1 Background

In spite of dynamic growth in the industrial and service sectors, growth in the agricultural sector has lagged behind. In particular, during the last two decades, there has been a distinct slow down in the growth in agricultural sector. As a result, the gap between the agricultural and other sectors has widened. Public investment has perceptibly declined, extension and other support services were largely inadequate and major crop yields have stagnated. Due to risks and uncertainties in production, marketing systems, access to institutional credit by the small and marginal farmers have been very low. The other difficulties faced by the small farmers are securing adequate collateral for getting bank loan, increased expenditure on crop cultivation, marketing bottlenecks, and low returns. Therefore, the typical farming in eastern India is all about poor farmers, fragmented landholdings, minimum farm mechanization, traditional agricultural practices, low use of inputs, and poor market linkage. All these have resulted in poor performance of the sector in spite of robust overall economic growth, leading to farmers continuing in a low investment and low return production cycle and increasing agrarian distress, which are manifested through forced migration, farmers' suicides, and reduced income from agricultural activities. Poor agricultural growth in eastern India has also hampered the country's food security.

Recognizing that the gap between available potentials and actual growth in the agricultural sector is maximum in the eastern region of the country, the Government of India, in the Union Budget for 2010–2011, announced a package of ₹ 400 crore for enhancing agricultural production in eastern India. To facilitate growth in agriculture, arresting poverty and enhancing income levels of farmers, an attempt has

---

K. G. Karmakar (✉)
S.P. Jain Institute of Management & Research, Munshi Nagar, Dadabhai Road, Andheri (W), Mumbai, Maharashtra 400058, India
e-mail: karmakar@spjimr.org

B. B. Sahoo
National Bank for Agriculture and Rural Development (NABARD), Bangalore, Karnataka, India

M. Ghosh et al. (eds.), *Diversification of Agriculture in Eastern India*,
India Studies in Business and Economics, DOI 10.1007/978-81-322-1997-2_2

been made to draw up an Action Plan for effective utilization of the newly created fund and enhancing crop productivity.

2.2 National Challenges

Indian agriculture has been passing through a difficult phase. While the rate of growth in agriculture has been decelerating, the absolute number of population dependent on agriculture has been increasing. Land degradation is widespread and fertile farmlands are being diverted towards non-agricultural uses. Due to increasing pressure on water and land resources from other sectors, the overall challenges are to ensure greater productivity with less water and less land. Agriculture production and farm incomes are frequently affected by natural disasters such as droughts, floods, pestilence, hail, cyclones, etc. Most agriculturally developed parts of the country are overexploiting their water resources, and possibilities for further expansion of irrigation are limited. The vulnerability of agricultural production to these disasters is compounded by the outbreak of epidemics. Global warming and climate changes are also posing threats to the stability of farm production.

Indian agriculture is dominated by small holders (85 %), who suffer from several production and marketing constraints. In spite of various reforms in agricultural marketing, linkages between producers and consumers remain weak, farm delivery services are poor, and value addition is very low. Linkages between the laboratory and the field have weakened. Involvement of the corporate sector in contract farming, value addition, processing, and marketing is very limited. Although there are a large number of farmers' club, self help groups, joint liability groups, and non-governmental organizations, they are not very active either in enhancing crop productivity or in offering necessary training to the farming community.

Rainfed areas, which are characterized by low levels of productivity and low input usage, constitute about 60 % of the gross cropped area (GCA) in the country. Secondly, most of the rural poor live in the rainfed regions. Further, stagnant farm output and rising population levels have resulted in a demand-supply mismatch leading to rising prices of essential commodities and dependence on imports. Along with enhanced use of inputs due to seed-fertilizer-irrigation technology, the energy inputs such as diesel, electricity, chemical fertilizers, etc., have also increased, resulting in an adverse ratio of crop output to energy inputs over time. Several factors such as diversion of cultivable land for commercial and bio-fuel crops and increase in cost of inputs like diesel oil and other sources of energy have resulted in skyrocketing prices, food crisis and has threatened food security, especially for the poor. The average farm size has been declining. Public investment in creating rural infrastructure has also been declining. Further, in many villages, sizeable cultivable areas are left fallow.

Experts opine that frequent occurrence of floods, drought, and irregular rain increase the financial burden of the common farmers. Rise in cost of cultivation, stagnant crop yields, and un-remunerative prices of agricultural produce have adversely affected the profitability of the farmers. As a result, many farmers are considering alternate occupations. The benefit of minimum support price (MSP) has been restricted to a handful of farmers producing notably wheat, rice, sugarcane, and cotton.

In spite of available arable land, water resources and fertile soil, food security in the country is in danger. The yield growth for many crops in the country has declined perceptibly in the 1990s. When the requirements of cereals, pulses, roots and tuber crops, milk, and oil and fats in the country are 450, 55, 63, 200, and 40 million t, respectively, the actual production of these food items are 216, 14, 31, 90, and 20 million t, respectively. These production gaps need to be met by enhancing productivity.

There exists a large yield gap in major crops produced in the country. It is evident from the 2003–2005 data of the Planning Commission that the yield gaps in wheat ranged between 6 % (Punjab) and 84 % (Madhya Pradesh); rice over 100 % in Assam, Bihar, Chhattisgarh, and Uttar Pradesh; maize between 7 % (Gujarat) and 300 % (Assam); jowar between 13 % (Madhya Pradesh) and 200 % (Karnataka); mustard between 5 % (Haryana) and 150 % (Chhattisgarh); soybean between 7 % (Rajasthan) and 185 % (Karnataka), and sugarcane between 16 % (Andhra Pradesh) and 167 % (Madhya Pradesh).

2.3 Methodology of the Analysis

Keeping in view the priority, the analysis has been restricted only to eastern India, i.e. five states (Bihar, Chhattisgarh, Jharkhand, Orissa, and West Bengal) and eastern districts of Uttar Pradesh (27 out of the total 71 districts in Uttar Pradesh). To avoid data limitations, all information regarding crop area, cropping pattern, input availability, and yield gap have been collected for the year 2005–2006. A total of 34 crops consisting of 5 cereal crops, 3 pulses, 9 oilseed crops, 3 fiber crops, 9 horticultural crops, and 5 other crops have been considered for the analysis. The data relating to crop area, cropping pattern, farm credit, and crop productivity were collected from secondary sources from Centre for Monitoring Indian Economy (CMIE), Economic Survey, Directorate of Economics and Statistics, and National Bank for Agriculture and Rural Development (NABARD). The ground level credit flow has been taken as a proxy for agricultural credit. Jowar, bajra, and maize have been considered under coarse cereals. In the absence of any information, sum total of net sown area and fallow land has been considered as the land available for crop cultivation. Further, net sown area has been taken as the kharif cultivated area. The Sen Committee Report, which looked into the reasons for low agricultural productivity in the mid-1980s, was also reviewed.

2.4 Agricultural Scenario: Eastern India vis-à-vis All India

The challenges before eastern India are not very difficult, from those at the all India level. In spite of good rainfall, abundant surface, and ground water resources and plentiful of labour, land use, crop diversification, crop yield, and farm income are not up to the mark. Rain water is not efficiently utilized due to inadequate provision of irrigation structures such as canals, gully-plugs, check dams, and irrigation projects. As a result, water resources in monsoon season are unevenly spread or controlled leading to floods and there is little water in dry seasons, leading to drought. In spite of several development programmes in the eastern region such as Command Area Development Programme, Drought Prone Area Development Programme, and Comprehensive Area Development Programme, most of the programmes fail to offer satisfactory results due to leakages, improper planning, and inadequate monitoring and control. While ground water exploitation in some states in the South and West have reached saturation level, utilization levels are only 8 % in Orissa, 24 % in West Bengal, 25 % in Bihar, and 32 % in eastern Uttar Pradesh.

2.4.1 Land use Pattern

Out of the total 166.07 million ha of the cultivable land in the country, 24.18 million ha (14.56 %) have remained fallow. When 82.62 million ha (42.86 % of GCA) is under irrigation, second crop is cultivated in 50.91 million ha (35.88 % of net sown area). However, in the eastern region, fallow land constitutes 14.21 % of the total cultivable area, and irrigation facility is available to 47.76 % of the GCA. When the cultivable land in the eastern region constitutes 22.32 % of the total cultivable land in the country, those of net sown area, rabi-cropped area, gross irrigated area, and cropping intensity in the eastern region are higher. However, among the eastern states, the maximum fallow land is found in Jharkhand (53.4 %) followed by Orissa (13.0 %). Similarly, the cropping intensity in the eastern region ranges between 102.74 % (eastern Uttar Pradesh) and 180.04 % (West Bengal). Table 2.1 presents the land use pattern in eastern India vis-a-vis all India.

The cropping pattern in the eastern India is quite similar with that in all India level. But when different states are arranged in a descending order from 1 to 10 in terms of area and productivity under different crops for the period between 1999 and 2000 and between 2005 and 2006, it is found that farmers of most of the eastern states have provided a higher share of their cropped area for the selected crops but the yield levels in these states are not very significant. Table 2.2 presents the rank of the selected five states, i.e. Bihar, Chhattisgarh, Jharkhand, Orissa, and West Bengal, in area and yield under different crops.

Table 2.1 Patterns of land use in the eastern region and all India during 2005–2006. (Source: Centre for Monitoring Indian Economy (CMIE), Various Issues, Directorate of Economics & Statistics, Various State Governments)

Sl. No.	Particulars	Eastern India	All India	Eastern region as a % of all India
1.	Cultivable land	37.07	166.07	22.32
2.	Net sown area	31.80	141.89	22.41
3.	Fallow land	5.27	24.18	21.79
4.	Fallow land to cultivable land (%)	14.21	14.56	–
5.	Rabi cropped area	12.07	50.90	23.71
6.	Net irrigated area	14.15	60.20	23.50
7.	Gross irrigated area	20.13	82.62	24.36
8.	Cropping intensity (%)	137.95	135.87	–
9.	Irrigated area to gross sown area (%)	47.46	42.86	–

Table 2.2 Rank of the selected states in eastern region by area and yield of different crops between 1999 and 2000 and between 2005 and 2006. (Source: Centre for Monitoring Indian Economy (CMIE), Various Issues, Directorate of Economics & Statistics, Various State Governments)

Sl. No.	Crop	Bihar	Chhattisgarh	Jharkhand	Orissa	West Bengal
1.	Rice	5 (6)	6 (10)	–	3 (7)	2 (2)
2.	Wheat	6 (6)	–	–	–	8 (7)
3.	Coarse cereals	8 (7)	–	–	10 (10)	–
4.	Pulses	7 (9)	10 (10)	–	9 (8)	–
5.	Groundnut	–	–	–	9 (9)	10 (10)
6.	Rapeseed and mustard	8 (8)	–	–	–	4 (4)
7.	Sesame	–	–	–	10 (–)	9 (3)
8.	Linseed	5 (3)	(4)	–(10)	7 (6)	8 (–)
9.	Castor	10 (8)	–		7 (6)	
10.	Niger seed	–	3 (5)	6 (3)	2 (1)	10 (10)
11.	Cotton	–	–	–	10 (10)	–
12.	Jute	2 (2)	–	8 (7)	5 (4)	1 (1)
13.	Vegetables, root, and tubers	–	–	9 (–)	2 (2)	1 (1)
14.	Potato	3 (3)	–	–	–	2 (2)
15.	Ginger	–	–	–	2 (4)	4 (5)

Figures in parentheses are rank of the states in yield

2.4.2 Agricultural Inputs

Water is a key ingredient in deciding cropping pattern and use of modern inputs including chemical fertilizers and pesticides for increasing the crop yield. The average rainfall received in the eastern region at 937.53 mm is 6.66 % more than that of all India level. But the average deviation from normal rainfall in the eastern region

Table 2.3 Details of annual rainfall, irrigation potential, fertilizer use and credit available during 2005–2006. (Source: Centre for Monitoring Indian Economy (CMIE), Various Issues, Directorate of Economics & Statistics, Various State Governments)

Sl. No.	Particulars	Eastern India	All India	Eastern region as a % of all India
1.	*Annual Rainfall (mm)*			
(a)	Actual	937.53	879.0	106.66
(b)	Normal	1120.86	893.0	125.52
(c)	Deviation from normal (%)	(−) 16.36	(−) 1.57	–
2.	*Irrigation potential (million ha)*			
(a)	Targeted	1.73	10.92	15.85
(b)	Exploited	0.83	5.16	16.18
(c)	To be exploited	51.71	52.71	–
3.	*Fertilizer consumption (kg/ha)*	121.27	105.50	114.95
4.	*Credit to farmers (Rs./ha)*	3946.87	9131.00	43.22

is (−) 16.36 % as against (−) 1.56 % at the all India level. The fertilizer consumption in the eastern region at 121.27 kg/ha is 14.95 % higher than that of 105.50 kg/ha at all India level. However, it is disturbing that in spite of a large number of financial institutions in the region and an array of financial reforms, credit availability per hectare in the eastern region is only ₹ 3946.87 as against the all India average of ₹ 9131.00. At the disaggregate level, the situation is quite alarming as the average credit available in Bihar, Chhattisgarh, and Jharkhand have not even reached to ₹ 3000 per ha during 2005–2006. Table 2.3 presents the rainfall, irrigation potential, fertilizer use, and credit availability in eastern region vis-a-vis all India during 2005–2006.

2.4.3 Crop Yields and Yield Differences

Table 2.4 presents the crop yields in the eastern region vis-a-vis all India during the year 2005–2006. In almost all crops, eastern region falls behind all India average in crop yield. For instance, when the areas under cereals, pulses, oilseeds, and sugarcane in the eastern region as a percentage of all India are 50.15, 10.54, 5.61, and 12.84, respectively, crop output of these crops as a percentage of all India are 27.32, 11.33, 4.16, and 10.37, respectively. Only in the case of potato production, the eastern region gets an edge over the all India average, i.e. 51.40 % of area but 54.65 % of output. It is disturbing to note that in spite of the dominance of the eastern region over the all India scenario in terms of cropping intensity, irrigation, rainfall, and fertilizer use, crop yields of rice, wheat, oilseeds, sugarcane, ginger, turmeric, and garlic are lower than all India average by 15.65, 34.15, 25.93, 19.20, 48.01, 55.73, and 34.65 %, respectively.

Table 2.4 Yields of selected crops in the eastern region and all India during 2005–2006 (q/ha). (Source: Centre for Monitoring Indian Economy (CMIE), Various Issues, Directorate of Economics & Statistics, Various State Governments)

Sl. No.	Particulars	Eastern India	All India	Eastern region as a % of all India
I. Food grain crops				
(i)	Cereals	17.75	19.92	89.12
(ii)	Rice	17.74	21.03	84.35
(iii)	Wheat	17.25	26.19	65.85
(iv)	Coarse cereals	19.97	11.62	171.83
(v)	Pulses	6.43	5.98	107.49
II. Non-food grain crops				
(i)	Oilseeds	7.44	10.04	74.07
(ii)	Sugarcane	540.81	669.28	80.80
(iii)	Potato	197.68	185.92	106.33
(iv)	Ginger	18.39	35.37	51.99
(v)	Turmeric	21.92	49.52	44.27
(vi)	Garlic	28.98	44.34	65.35

2.5 Action Plan

2.5.1 *Strategy*

2.5.1.1 Expansion of Crop Land

By increasing net sown area and enhancing double-cropped area, the GCA can be increased. Net sown area can be increased by planting bamboo and tree crops such as mango in fallow lands and wastelands. Double crop area can be increased by increasing irrigation facilities, saving monsoon water through rain water harvesting structures, using sprinkler irrigation system, and using suitable kharif area for seasonal vegetable, roots, and tuber crops. The rain harvesting models of Gujarat and sprinkler/drip irrigation models of Andhra Pradesh could also be emulated.

2.5.1.2 Crop Diversification

When the cropping pattern in eastern region vis-à-vis all India scenario is observed during the last one and a half decades, it is found that cropping pattern in the eastern region has hardly changed over the last one and a half decades. This is evident from the fact that food grain crops constitute 78.83 % of the total cropped area of 41.41 million ha in the eastern region as against 63.07 % (192.80 million ha) at the all India level. It is also a fact that repeated production of same crop from a piece of land degrades soil quality and reduces crop yield. Therefore, to keep the soil healthy and improve crop yield besides increasing farm profitability, the cropping pattern in the eastern region must change. Location-specific crop-mix in accordance with soil, climate, and rainfall and irrigation facility needs to be put in place. To improve the food security situation, farmers may be incentivized for the cultivation of non-food grain crops such as oilseeds, fibre crops, horticulture crops, vegetables, onion, and potato. Pulses, the principal source of protein for a vegetarian diet, face challenges in raising production levels. The availability of pulses has also been steadily declining. In dry lands, different pulses such as green gram, black gram, and horse gram may be tried. Innovative models such as Pulses Village, Oilseed Village, and Seed Village may be developed. These innovations will enhance the financial margins of farmers in the eastern region. Small/marginal farmers could also diversify into dairy/animal husbandry activities and fish rearing in ponds, which would also help in de-risking their agricultural operations.

2.5.1.3 Rural Infrastructure

Infrastructure is the key to agricultural progress. The Government has been playing a big role in creating core infrastructure such as major and medium irrigation projects, rural roads, rural electrification, and setting up of agricultural markets. Acknowledging the sharp decline in public investment in agriculture during the 1980s, the central and state Governments have been taking steps to reverse the declining trend. It is proposed to raise the level of public investments to 4 % of gross domestic product (GDP) agriculture by the end of the 11th Five-Year plan, i.e. by the year 2011–2012. In spite of the efforts of the Government in building rural infrastructure through a host of schemes such as Rural Infrastructure Development Fund (RIDF), Accelerated Irrigation Benefit Programme (AIBP) and Bharat Nirman, the irrigation potential created, availability of quality seeds, fertilizers, credit, and extension facilities and agricultural marketing in entire eastern region as against the all India situation, are not comparable. To improve the situation, restore, and maintain the created infrastructure, public participation is urgently needed. Involvement of the private sectors in building infrastructure, maintaining, monitoring, and evaluating the impact of the created infrastructure, building value chains, disseminating technical knowledge among the farmers groups such as self-help groups and farmers' clubs and creating a conducive atmosphere for public private partnership, may be encouraged through a suitable policy framework.

2.5.1.4 R&D and Use of Technology

The main challenge in research & development related to agriculture has been to improve the linkage between research and extension and to reform the system to improve the relevance of such institutes. Innovation of heat and drought resistant varieties of seeds, pest management, and manuals on easy and understandable farm practices should be brought out and widely disseminated through lab-to-land experiments so that farmers can emulate such practices. One of the research priorities should be evolving strategies to increase yields and incomes of the small farmers in rainfed conditions with the help of low cost and low technologies methods. On a selective basis, in upland areas, systemic rice intensification (SRI) methods may be applied as also systemic sugarcane intensification (SSI) methods so as to conserve water and input use. Besides, the Prof. Dabholkar Model of increasing crop yields as per Natueco Organic methods may be tried.

2.5.1.5 Reducing Risk in Agriculture

Agriculture has two types of risk, i.e. Yield risk and Price risk. As the major cultivated area is dependent on rainfall, crop insurance is important for farmers. To cater to the needs of farmers, the Government has introduced the National Agricultural Insurance Scheme (NAIS) from the Rabi 1999–2000 season. However, certain shortcomings relating to calculation of generated income, low indemnity levels and delays in settlement of insurance have been observed in the implementation of the scheme. Further, crop insurance is not a long-term solution for yield variability. We need to focus on land and water management techniques including irrigation developments in the public delivery system. To address price risks, commodity futures markets are advocated. However, it is not clear whether farmers are actually benefiting from futures markets. For de-risking price risks and yield risks, contract farming and area-based rainfall index insurance may be thought of Banks will be able to meet production credit, investment credit, and consumption credit needs of the farmers with a consolidated upper limit. The present Kisan Credit Card (KCC) is actually being used as only a passbook for recording transaction entries. It should be designed for operations over a 5 years period, as a genuine credit card over a 5 years cropping period, which is adequate for a crop pattern of 2 average years, 2 good years, and 1 of below average crops. At the end of the 5-year period, dues could be re-scheduled in un-irrigated areas, depending upon the reasons for non-settlement of bank dues.

2.5.2 Emerging Action

We need to act on many fronts for increasing crop productivity. Among others, research and innovation hold the key. Agriculture Universities and Research Institutes should enhance their research work and put in place a system by which lab-to-land exchange of knowledge is done. Biotechnology is the key to improving crop yields. Sharing and emulating noteworthy best practices and exemplary case studies across Krishi Vigyan Kendras (KVKs) would usher in a modernizing influence on agricultural production. As most of the farmers in the eastern region do not have access to formal sources of credit, necessary steps may be taken by the government and financial institutions. In this connection, the Lead Bank Scheme (LBS), introduced in 1969, needs to be reviewed. The RBI review of the LBS in 2010 was a shoddy attempt at maintaining the status-quo. In the context of rising food prices, reviving agriculture production, and productivity is crucial. Action is needed for better management of water resources, quality seeds availability, adoption of scientific farm practices, rational use of fertilizers, integrated pest management, and diversification of agriculture. The key to growth for productivity is adoption of new technologies, especially biotechnology. Bt. Cotton is a shining example of how biotechnology can revolutionize production. In 5 years of introduction of Bt. cotton, productivity of cotton has almost doubled. When rice productivity in India is below 3 t/ha, China is developing super hybrid rice, targeting yield levels of 15 t/ha. As these factors are all inter-related, a co-ordinated R&D approach is necessary.

The eastern region is dominated by small and marginal farmers, and most of these farmers sell their produce at below MSP. Effective marketing support may be provided to farmers so that the gap between producers' prices and consumers' prices can be narrowed down. In drought prone areas, low cost irrigation has to be put in place and drought tolerant vegetables and crops such as papaya and banana plantations apart from nursery and tree plantation may be experimented with. Research needs to focus on crops and cropping systems in the dry lands, hills, tribal, and marginal areas. Dry land farming technology has to be improved. Further, research has to be increasingly location-specific with greater participation of farmers. Horticulture crops that are land and water saving should be encouraged in dry land areas. Wastelands may be used for bamboo and other tree plantations.

In the absence of suitable marketing facilities in the region, most of the farmers have been selling their surplus produce at un-remunerative prices soon after harvests. Adequate strong facilities/spaces may be created in rural areas through public–private partnership to provide price advantage to the farmers. The extension machinery should make use of the available infrastructural facilities such as rural godowns, cold storages, and watershed scheme. Studies show that establishment of rural godowns have changed the agricultural scenario in items of change in cropping pattern, employment, marketing of produce, and enhancing farm income. Similarly, due to watershed projects, the farmers in the catchment areas derive a variety of benefits such as increased cropping area, increased functional efficiency and regulated irrigation, stabilised crop production, prevention of soil erosion and sedimentation of water bodies, and increased employment and income.

Experts opine that dry lands are best suited for organic agricultural practices. The climate, growing conditions, untapped production potential, vast human resources, least use of chemical fertilizers/pesticides, and a strong traditional farming system are some other advantages for such crop production. The potential areas for the growth of organic farming are rain-fed, tribal, and hilly regions. To popularize organic practices and reap the benefits of organic farming, experts should investigate how the small and marginal farmers could benefit from such farming, and how dry and wastelands could be used for such farming. In this context, public–private partnership models need to be firmed-up. There exists a large knowledge gap between the yields in research stations and actual yields in farmers' fields. Agri-clinics and agri-business centres may be established through creation of technology agents to provide vocational training so that effective translation of lab-to-land experiments can be encouraged. Groups of farmers may be encouraged to form "producers' companies" and make all arrangements for buying inputs and selling their produce at remunerative prices.

Collaboration with agricultural universities and research institutes is necessary for utilizing the latest research findings on high yielding seeds, ideal cropping pattern, and effective skills-up-gradation of the members of groups such as farmers' clubs, self-help groups, and joint liability groups. In order to integrate farmers with input markets, output markets, and supply chains, appropriate infrastructure, sustainable production intensification, innovative marketing models, and value chains may be considered. An enabling environment may be created to attract corporate sectors for effective public–private partnership by putting in place proper policies and incentives. As imposition of any new system on farmers is not an easy task, it is important to make use of local experts as the change agents. Further, the participation of farmers' clubs, self-help groups, joint liability groups, and local non-governmental organizations will improve the situation. The members of the NGOs, farmers' clubs, self-help groups, joint liability groups, and progressive farmers should also be properly trained to act as change agents.

Harnessing Information and Communications Technology (ICT) is crucial for disseminating knowledge, providing facilities for soil, nutrient and water testing facilities, marketing opportunities, and sharing experiences of best practices. For providing appropriate agricultural extension services, there is a need to stabilize the Common Service Centres (CSC). As a result of collaboration between NABARD and Indian Council of Agricultural Research (ICAR), the first e-village in the country has been set up at Saiden, Mizoram. At Saiden, ICT is being used for testing soil health, predicting rainfall, detecting, and assessing the decline in productivity of citrus fruit. Similarly, the issue of soil health cards has increased agricultural productivity in Gujarat. The e-Sagu experiment of Dr. Krishna Reddy of IIIT, Hyderabad, the e-Choupal network by ITC, Secunderabad and the e-Kutir Pilots in Orissa by Intel/Grameen Bank have succeeded in improving the agricultural productivity by providing market information, creating awareness on quality agricultural information, and promoting sustainable use of inputs such as seeds, fertilizers, pesticides, and water. Thus, agri-extension services using ICT have already been field-tested. Therefore, the best practices of ICT-led agricultural growth through active involvement of NGOs along with KVKs and Agriculture Technology Management Agencies (ATMAs) may be tried on a pilot basis in all districts of the eastern region, with suitable scale-up within the districts, later. Use of mobile phones to disseminate information to farmers on inputs, weather and commodity market prices would be cost-effective.

2.6 Conclusions

The eastern region, in comparison with all India agricultural production figures, is backward in crop diversification and crop productivity. Infrastructure, credit flow, marketing facilities, insurance available, price structure, and information flow provide little support to the farmers to increase their sown area, improve productivity, and increase farm incomes. In spite of abundant arable land, water, bio-diversity, and large population, when more than half of the total rice area in the country is located in the eastern region, the region provides only 42.30 % of the total rice output. The yield of rice in the region is 1773.91 kg/ha as against all India average of 2103 kg/ha. Therefore, the main challenge in the region is to enhance the productivity of rice and rice-based systems. Keeping in view the vulnerability of the region to multiple risks of floods, drought and cyclones, etc., genetic improvement in rice variety must be adopted for higher yield and stability and better crop management. The systemic rice intensification (SRI) method being used in Andhra Pradesh, Tamil Nadu, and Tripura, could be utilized in upland areas of all eastern region states so as to improve yields and reduce input costs by over 40 %. The focus of the R&D could be to prioritize the rice research in the eastern region to reduce the yield gap. By extrapolating, it can be shown that when the yields of rice and wheat in the region reach the all India average level, the country's total cereals production increases by 12.58 million t. However, in a global framework, when demand for farm produce is increasing and the land in the eastern region is suitable for production of a variety of crops including fruits and vegetables, the value chain system partnered with the corporate sector and support from the financial sector, can succeed in bringing about all round improvement. To make this happen, a suitable policy framework has to be put in place. Unless there is determination to address issues vigorously, enhancing production, and productivity in eastern India will remain a myth and will continue to affect food security for the entire country, adversely.

Chapter 3
Diversified Sustainable Agriculture in Eastern India

Tapas Kumar Chakrabarty

3.1 Introduction

Farmers, specially marginal and small, particularly in eastern India, still face multidimensional distress. The deficiencies in institutional factors like credit, insurance, supply of inputs and marketing have become serious. Social factors such as the exploitative attitude of moneylenders cum merchants and the lack of sound collaborative forward mindset have aggravated farmers' distress (Anonymous 2007; Swaminathan 2004).

Agricultural diversification is a strategy to overcome economic distress of the rural sector of any economy. Diversification in agriculture generally refers to diversification between agriculture and allied activities like animal husbandry, fishery, etc. (vertical diversification) and diversification in cropping pattern (horizontal diversification; Haque 2010).

The global demands as well as the changing domestic consumption pattern has provided opportunities for expanding the range of agricultural products and allied products (Basu 2001). However, farmers need to be properly knowledgeable and market-oriented. Trading in futures, contract farming along with other available facilities may provide inspiration to farmers willing to adjust for their own wellbeing first, and then for others through sound and purposeful collaborations. Under this background, this chapter attempts to discuss the issues of mindset of the farming class in eastern India.

T. K. Chakrabarty (✉)
'Purbachal', Andrews Pally, Santiniketan - 731235, Birbhum, West Bengal, India
e-mail: chakrabartytk@gmail.com

Formerly, in Reserve Bank of India, Mumbai, India

M. Ghosh et al. (eds.), *Diversification of Agriculture in Eastern India,*
India Studies in Business and Economics, DOI 10.1007/978-81-322-1997-2_3

3.2 Recent Agricultural Situation

As is well known, the year 1968 marked the beginning of a turning point in Indian agriculture. The development of a high-yielding variety (HYV) of seeds in the mid-1960s and the subsequent use of the fertilizer–pesticides–irrigation package, better field machines and instruments, bank finance, conducive extension service, etc., led to quantum jumps in the productivity of some agricultural products, mainly wheat, rice and other food grains. The production of wheat, rice and food grains during the 1970s and 1980s grew at the annual average rate of 5, 4 and 3 %, respectively. The yields of these products grew at the annual average rate of 3 % each. The food grains growth (around 3 %) was higher than the population growth (around 2.2 %) during the 1970s and 1980s.

However, agricultural productivity in India has witnessed a fatigue since the early 1990s. The annual average growth rate of the production of wheat, rice and food grains between 1990 and 2010 came down to 3, 1 and 2 %, respectively. Yield of wheat, rice and food grains grew at the annual average rate of 2, 1 and 1 %, respectively during 1990–2010. However, the average population growth was around 2 % during 1990–2010. Consequently, the per capita net availability of food grains per day came down from around 510 g in 1991 to around 444 g in 2009 (Chakrabarty 2011a, 2011b; Gokarn 2011; Mohanty 2011a, 2011b).

It might be relevant to indicate that there exists a wide variation in yield of these crops across states/regions in our country. It can be seen from Table 3.1 that eastern-zone states experienced less than the all-India average productivity in major crops, and the all-India average yields of the crops was found to be less than that in China and the USA.

Low yield across major crops has become a regular feature of Indian agriculture in recent years. This can be attributed to the structural weakness of the agriculture sector reflected in low level of public investment, exhaustion of the yield potential of new high-yielding varieties of wheat and rice, unbalanced fertilizer use, low seed replacement rate, inadequate post-harvest value addition, price volatility, etc.

Eastern India faced some diversified agricultural development over the years (Haque 2010; Karmakar and Sarkar 2011; Ghosh 2011). There was also a structural transformation in the economy. The contribution of agriculture and allied activities to gross state domestic product (GSDP) followed the declining national trend. The recent Indian growth is service-led. Agriculture provides around 60 % of employment, but its share in the gross domestic product (GDP) is currently around 14 %, fell from historical 50 % over the decades.

The contribution of agriculture and allied activities to GSDP is currently at around 25 % in Bihar, followed by Orissa (23 %), West Bengal (21 %) and Jharkhand (10 %). Crop diversification in the eastern zone has been in favour of oilseeds, horticultural crops, vegetables, away from food grains (cereals, pulses) over the years. Livestock gained some importance, particularly in Bihar. There is a need for farmers to emphasize the improved activities towards horticulture, livestock, vegetables, oilseeds, etc., following the all-India food demand projection as shown in Table 3.2.

Table 3.1 Cropwise yield across states: 2010–2011. (Source: Address by Dr. K. C. Chakrabarty, Deputy Governor RBI at the National Seminar on Productivity in Agriculture at CAB, Pune on 2 Sept 2011)

State	Rank	Yield (t/ha)
Rice		
Punjab	1	4.0
Tamil Nadu	2	3.0
Andhra Pradesh	3	2.9
Haryana	4	2.8
Karnataka	5	2.6
West Bengal	6	2.6
Orissa	14	1.4
Jharkhand	19	1.2
Bihar	20	1.0
National average	–	2.2
China	–	6.7
USA	–	7.5
World average	–	4.3
Wheat		
Punjab	1	4.5
Haryana	2	4.4
Uttar Pradesh	3	3.2
Gujarat	4	3.1
Rajasthan	5	2.9
West Bengal	6	2.6
Bihar	8	2.0
Jharkhand	12	1.5
Orissa	13	1.4
National average	–	2.9
China	–	4.7
USA	–	3.1
World average	–	2.9
Oilseeds		
Tamil Nadu	1	2.2
Haryana	2	1.8
Maharashtra	3	1.4
Gujarat	4	1.3
Punjab	5	1.2
West Bengal	8	1.1
Bihar	11	1.0
Orissa	17	0.7
Jharkhand	20	0.4
National average	–	1.2
China	–	2.1
USA	–	2.7

Table 3.2 Demand projection (all India, million tonnes). (Source: Estimates by the Indian Council of Agricultural Research (ICAR))

Food items	Domestic production (2006–2007)	Demand projection (2020–2021)
Cereals	202	262
Pulses	14	19
Food grains	216	281
Oilseeds	24	54
Vegetables	112	127
Fruits	58	86
Fish	7	11
Milk	2.5	5
Egg (billion)	51	81

Estimates by the Indian Council of Agricultural Research (ICAR) show that in order to meet demand requirements of food items, namely oilseeds, fish, eggs and fruits by 2020–2021, the production has to grow at an annual rate of 6.0, 3.5, 3.4 and 2.9 %, respectively between 2006–2007 and 2020–2021 (Table 3.2). Initiatives from the farming community of eastern zone may get inspiration from the current market-oriented socio-economic and political will, exhibited by other agriculturally developed states.

3.3 Agricultural Marketing Facilities

Post-harvest value addition is the goal of the farming community to sustain their comfortable living. The precise causes of farmers' distress are many and varied. Market collapse and low prices for farm commodities are major factors in many places of rural India. Besides, traditional harassment by private moneylenders cum input merchants is still causing serious hardship to marginal and small farmers.

The absence of producer-oriented marketing systems in many places of rural economy is causing misery to the economically and socially underprivileged. If the market is less ill intermediated, the scope for better value addition to farmers is more. In order to reduce the exploitation of non-farming middle men like money lenders, bulk land owners, etc., agricultural markets are generally regulated before being developed in the private and cooperative sector.

Farmers over the experience should be able to undertake a market-driven production plan and adopt modern marketing practices. The liberalized agricultural market facilitates industries and large trading companies to undertake procurement of agricultural commodities directly from the farmers' fields and helps the post-harvest value addition. The importance of contract farming emerges. Price realization by farmers becomes better through the elimination of intermediaries.

During the Green Revolution, a major reform was initiated, when almost all the states brought in legislation the Agriculture Produce Market Committee (APMC)

Act. Around 8000 regulated markets under APMCs throughout the country perform the crucial function of organizing principal agricultural commodity wholesale trade and providing a meeting point for buyers and sellers for party-to-party cash or spot contract for ready delivery.

The geographical area in each state is divided in command areas for the regulated marketplace, wherein the market is managed by the market committee, constituted by the state government. All the principal agricultural commodities in India are traded in wholesale markets by free market forces of demand and supply. The government of India sets a minimum support price (MSP) for principal commodities (determined upon the fraction of total cropped area in the country occupied by cultivation of the commodity). Rice is the largest principal commodity since it has the largest fraction of total area under cultivation (say 50 %). There are around 25 agricultural commodities for which the government of India sets an MSP, which is considered as a threshold price in the market.

The development of regulated market yard infrastructure according to the modes of the sellers (primary farmers from the command area, secondary farmers and traders beyond the command area also, terminal traders only) under APMC Act is very lopsided, and its progress is satisfactory in some states like Rajasthan, Gujarat, Maharashtra, Karnataka, Punjab and Haryana.

The country has around 27,000 unregulated rural periodical markets. In the eastern zone, unregulated markets are playing a crucial role, due to vested local intermediation trapping marginal farmers. Marginal farmers are like bonded farmers. Socio-political effective will matters a lot. It may be argued that in each agro rich state, farming community plays a direct crucial positive role in the state and central politics to help rural sector. The subject of agriculture and agricultural marketing is dealt with both by the states and the central government.

Starting from 1951, the different 5-year plans laid stress on the development of physical markets, farm and off-farm storage structures, facilities for standardization and grading, packaging, transportation, etc. In 1965, the Central Warehousing Corporation (CWC), the Food Corporation of India (FCI), the Agricultural Prices Commission (later renamed as Commission for Agriculture Costs and Prices) came into existence. Besides, a number of organizations were set up in the form of commodity boards and cooperatives.

Federation and export promotion councils for monitoring and boosting the production, consumption, marketing and export of various agricultural commodities were set up. The prominent ones among them included the Cotton Corporation of India Ltd. (CCI), the Jute Corporation of India Ltd. (JCI), the National Cooperative Development Corporation Ltd. (NCDC), the National Agricultural Cooperative Marketing Federation Ltd. (NAFED), the Tribal Cooperative Marketing Development Federation Ltd. (TRIFED), the National Consumers Cooperative Federation Ltd. (NCCF), etc., for procurement and distribution of commodities. Commodity-wise initiatives like Tea Board, Coffee Board, Coir Board, Rubber Board, Tobacco Board, Spices Board, Coconut Board, Central Silk Board, National Dairy Development Board (NDDB), National Horticulture Board (NHB), State Trading Corporation (STC), Agricultural and Processed Foods Export Development Authority

(APEDA), Marine Products Export Development Authority (MPEDA), Indian Silk Export Promotion Council (ISEPC), Cashewnuts Export Promotion Council of India (CEPC), etc., are for promotion of production and export of specific commodities. Trade in commodities can be transacted in the spot directly as well as through registered brokers in the futures exchanges.

As there is a considerable time gap between the initial spending on seeds, fertilizers, pesticides, water, labour, etc., on the one hand and revenue from harvest on the other, farmers are vulnerable to commodity price fluctuations. Historically, the prices of agricultural products are volatile due to unpredictable weather conditions. This has made farmers look for alternatives to reduce risk. Buyers also think for avoiding price rise uncertainty. Futures trade/contract enables farmers to sell the crop at a specific price (price discovery) for delivery on the future date (risk management). Clearing houses of commodity exchanges guarantee the execution of these derivative instruments so that farmers get the price of commodity traded in futures.

If the price available in the futures market is not remunerative enough to the farmers, then he can change his cropping plan right at the beginning of the season. Due to small holdings of majority of farmers and a high level of illiteracy, farmer cooperatives can play a major role in aiding groups of farmers to mitigate price risk.

The futures market through exchanges facilitates buying and selling of standardized contractual agreements (for future delivery) of the underlying asset as the specific commodity and not the physical commodity itself (derivative). Futures trading helps price discovery and price risk management for the industrial and farming communities.

Forward/futures markets have come a long way since the days of the rice tickets in Japan and the first organized futures market in the form of the Chicago Board of Trade in the USA. In India, organized commodity derivatives trading began with the Cotton Trade Association's debut in futures in 1875. Cotton merchants of Bombay took cues from the USA and the UK, and to regulate futures trading, the government in 1918 set up the Cotton Contracts Committee, which was soon (1919) replaced by the Cotton Contract Board. Futures trading in oilseeds was organized with the setting up of Gujrati Vyapari Mandali in 1900 in Bombay. Over the years, the derivatives market developed in several other commodities in the country: raw jute and jute goods in Calcutta (1912), wheat in Hapur (1913) and bullion in Bombay (1920). During World War II (1939–1945), the world economy faced depression. Indian financial markets were put under control. Consequently, the Indian commodity futures market slipped into virtual extinction with some negligible deal in the form of over the counter.

With the process of liberalization and globalization of the Indian economy and consequent reforms in its financial markets in the early 1990s, the Prof. K. N. Kabra Committee, set up by the government in 1993 to examine the role of futures trading, made several recommendations including certain amendments to the Forward Contracts (Regulation) Act 1952 and the strengthening of the Forward Markets Commission (FMC).

Table 3.3 Ranking of states as per futures on potato in 2009

State (21 no)	Ranking
New Delhi	1
Uttar Pradesh	2
Maharashtra	3
West Bengal	4
Bihar	17
Jharkhand	18
Orissa	20

The Government allowed future trading in all the commodities recommended. The trade came into being after remaining in hibernation for nearly four decades, as the realization that derivatives do perform a role in risk management dawned. As the 1990s had an upswing in the global commodity cycle, FMC and the government acted urgently on setting up commodity exchanges with state-of-the-art infrastructure and global best practices. Three national-level on-line exchanges—the Multi Commodity Exchange of India Ltd. (MCX), the National Commodity and Derivatives Exchange Ltd. (NCDEX) and the National Multi Commodity Exchange Ltd. (NMCE)—were born to integrate the asymmetrically informed or ill-informed, fragmented domestic physical markets through effective application of information, communication and technology (ICT) for being a price setter, not price taker. Besides, around 21 commodity-specific regional futures exchanges are operational. Since the reintroduction of commodity futures trading in India in 2003, the bulk of trading has been taking place on the three national exchanges. Despite being a late starter, MCX overtook other domestic exchanges and continues to be No. 1 commodity futures exchange in the country (by numbers/lots of contracts traded) with a market share of 85 % (as on 31 August 2009). Going through some latest available data from MCX, the reflection comes that the eastern zone's involvement in trading in organized agromarkets, particularly in futures is not significant (Table 3.3).

Small-scale farming may be the main hurdle for the eastern-zone farmers to reap the benefits of modern organized marketing facility. Besides, small-scale farmers are virtually bonded due to low and irregular post-harvest value addition. In 2004, the State governments and the representatives of trade and industries at the National Conference of State Agriculture Ministers agreed to take steps in establishing an institutional set up for supporting contract farming.

The main feature of contract farming is that farmers grow selected crops under a buyback agreement with an agency engaged in trading or processing. There are many success stories of contract farming, such as potato, tomato in Punjab, sunflower in Madhya Pradesh, oil palm in Andhra Pradesh, seed production for hybrid seed companies in Karnataka, cotton in Tamil Nadu and Maharashtra. The contractual agreement with the farmer provides access to production services, credit and knowledge of new technology. Pricing arrangements can significantly reduce the risk and uncertainty of the market place (Anonymous 2007; Chakrabarty and Shunmugam 2009; Chakrabarty and Giridhar 2007).

3.4 Conclusions

Instead of lamenting that productivity has been low, one must look at the past to the Green Revolution in the mid-1960s to draw confidence for revival. States in the eastern zone are characterized by diversity of climate, soil and topographical conditions allowing for cultivation of varieties of crops. The nature has provided this region with ample inputs for the production of varieties of agricultural crops. Farmers' mindset needs to be led towards the knowledge-based thinking like that of farmers of agriculturally developed states of the country. The problems of marginal and small-scale farmers of the eastern zone can be addressed through the state initiatives supporting contract farming.

Farmers using the current demographic dividend need to be effectively and spontaneously cooperative in agro-related economic activities. The cooperative venture will motivate the farming community to gather knowledge for usefulness of modern production, marketing facilities and other facilities for regular post-harvest value addition.

In this regard, a transparent social and political will is a must, so that rural extension service will be effectively utilized for rural life comforts. A folk tale from Rajasthan suggests 'Concentrate on main activity, branching out to irrelevant activities will be not only troublesome, it may defeat the very purpose as well' (Dadhich 2011).

Farmers of the eastern zone must concentrate on using modern economic ways and means available only to improve their economic lot first, then linkage effects to others. Contract farming and trading in futures deserve attention. Finally, the following complementary comments may be relevant.

Crop diversification in eastern zone was experimented towards oilseeds, horticultural crops, vegetables, away from food grains (cereals and pulses) over the last decade or so. Livestock also gained importance over the years particularly in Bihar. In this direction, sound and purposeful emphasis by the farmers through apolitical collaborative team is the demand of the time to explore the demand-led opportunities.

The demand projection estimate suggests that in order to meet demand requirements of food items like oilseeds, fish, eggs and fruits by 2020–2021, production needs to grow at an annual rate of 6, 3.5, 3.4 and 2.9 %, respectively during the interval years. Indian households have begun to diversify their diets away from predominantly cereals to protein sources—pulses, milk, eggs, meat and fish—vegetables and fruits. In fact, the recently published Consumer Expenditure Survey carried out by the National Sample Survey Organisation in 2009–2010 demonstrates how significant the shift from cereals to proteins has been in terms of its significance in household food budgets.

The country went through a Green Revolution in the late 1960s and early 1970s, during which cereal production was raised dramatically to satisfy demand of rising population. However, recently the gap between rapidly rising demand and not so rapidly rising supply of proteins, vegetables and fruits has emerged. The simple reason for this is the demographic dividend, contributing to non-rural growth and productivity.

As per a United Nations projection, a high-consumption cohort in India's total population will continue to dominate the demand till 2040. With the gradual increase in the popularity of the APMC market initiatives in the eastern zone, farmers, capitalizing the demographic dividend (youth dominance) need to change the mindset to embrace modern technical, marketing, financial and insurance facilities for a fruitful diversification of agricultural activities. Small farms call for sound and purposeful collaboration.

The political will holds the key. Panchayati Raj institutions need to be involved more proactively to inspire a sound collaborative market-oriented mindset of rural young Bengal. The role of agricultural research institutions and related support agencies cannot be lost sight of launching trade, finance and insurance literacy. They are to ensure that there is a supply of appropriate and affordable services available to those who need them.

References

Anonymous (2007) MCX certified commodity professional (MCCP) reference material. MCX, Mumbai

Basu RL (2001) Glimpses of Indian economy: ancient and modern. Smashwords, Calcutta

Chakrabarty KC (2011a) Agricultural productivity and credit—issues and way forward. RBI Mon Bull, Oct

Chakrabarty KC (2011b) Financial inclusion: a road India needs to travel. RBI Mon Bull, Nov

Chakrabarty C, Giridhar VV (2007) Reaping rich dividends: trading in futures helps farmers in insuring prices of their produce against fluctuations. Hindustan Times Delhi, 13 December 2007

Chakrabarty C, Shunmugam V (2009) Commodity insights yearbook. PwC-MCX, Mumbai

Dadhich CL (2011) Management through folk wisdom. Diamond Books, New Delhi

Ghosh JK (2011) Status of agro-based industries in West Bengal: constraints and tasks ahead. Paper presented in the national seminar on prospects and challenges for rural economic development, organised by the Department of Economics and Politics, Visva-Bharati, Santiniketan, March 2011

Gokarn S (2011) Gokarn monetary policy: key factors shaping trajectory. RBI Mon Bull, Nov

Haque T (2010) Constraints and potentials of diversified agricultural development in eastern council for social development and others. Project report, Planning Commission, Government of India, New Delhi

Karmakar S, Sarkar D (2011) Study on regional pattern of agricultural growth in West Bengal. Paper presented in the national seminar on prospects and challenges for rural economic develpoment, organised by Department of Economics and Politics, Visva-Bharati, Santiniketan, March 2011

Mohanty D (2011a) Indian economy: progress and prospects. RBI Mon Bull, Oct

Mohanty D (2011b) Inflation: decoding the dynamics. RBI Mon Bull, Nov

Swaminathan MS (2004) Farmers' distress: causes and cures, economic developments in India–EDI 82. Academic Foundation, New Delhi

Chapter 4
Crop Diversification: An Exploratory Analysis

G. D. Banerjee and Sarda Banerjee

4.1 Introduction

Diversification at the macro level generally means that secondary and tertiary sectors become more important over time than the primary sector, in terms of their contributions to the national income as well as in disposition of the work force. There are three types of diversification viz. (a) a shift of labour from farm to non-farm activities, i.e. employment diversification; (b) a shift from a less profitable crop to a more profitable one, i.e. crop diversification; and, (c) use of resources in diverse but complementary activities, i.e. resource diversification. Diversification drew widespread attention in India in the recent past because of stagnant growth, incomplete agricultural transformation and low productivity. There are four dimensions of crop diversification: number of crops, spread of cropping pattern, proportion of high-value crop in the cropping pattern and shift in cropping pattern mix.

There is a need to address the issues of crop diversification that has been taking place in India, and to examine the link between different dimensions of diversification and the growth of output in India. The specific objectives of this paper are (1) to discuss the nature and extent of crop diversification across Indian states and to examine its role in the income and risk pattern in India, (2) to investigate the growth inductive or depressive impact of changing process of crop-mix and (3) to identify the factors influencing the typology of diversification in India in the last three decades.

G. D. Banerjee (✉)
Vivekanand Education Society's Institute of Management Studies and Research,
497 Collectors Colony, Chembur (E), Mumbai 400071, Maharashtra, India
e-mail: gandhar.banerjee@gmail.com

S. Banerjee
Mumbai, Maharashtra
e-mail: banerjee@gmail.com

M. Ghosh et al. (eds.), *Diversification of Agriculture in Eastern India*,
India Studies in Business and Economics, DOI 10.1007/978-81-322-1997-2_4

4.2 Concept and Rationale of Crop Diversification

Crop diversification is defined as the strategy of (i) shifting from less profitable to more profitable crops, (ii) changing of variety and cropping system, (iii) increasing exports and competitiveness in both domestic and international markets, (iv) protecting the environment and (v) making conditions favourable for combining agriculture–fishery–forestry–livestock. Crop diversification can be a useful means to increase crop output under different situations. It can be considered in two different ways, viz. (a) horizontal crop diversification and (b) vertical crop diversification.

The horizontal crop diversification refers to the addition of more crops to the existing cropping system. This means the broadening of the base of the system, simply by adding more crops to the existing cropping system utilising techniques such as, multiple cropping coupled with other efficient management practices. The vertical crop diversification means use of any crop species, which could be refined to manufacture products, such as fruits (canned or processed into juices or syrups). This will reflect the extent and stage of industrialisation of the crop. It also takes into account the economic returns from different crops. This is very different from the concept of multiple cropping in which the cropping in a given piece of land in a given period is taken into account.

Two major areas of concern of crop diversification are (a) availability of food for the ever-increasing population and (b) prevalence of poor condition of the people engaged in agriculture. To solve the above two problems there is a necessity to increase the volume of crop production. However, increase in crop production by enhancing the extent of agricultural land has become an impossible task. The other alternative is to bring additional land under cultivation. Hence, the only available option is to increase the intensity of cultivation and improvement of yields per unit area to meet future food requirement to feed an ever-increasing population, and improving the conditions of the poor farmers.

There are about 150 crop species, which have already received attention from breeders. However, only about 10–15 crops are produced in half of the world's food production. Novel technologies in emerging crops will facilitate farmers to stabilise their specific agro-environments, particularly by establishing suitable species and high-value cultivars with food, feed, fuel, fiber and pharmacological potential. The identification of alternative crops and improved technologies should aim at providing a comparative advantage within a given agro-ecological and socio-economic context. Sustainable intensification of agriculture without further degradation of natural resources remains a challenge.

Risk reduction through diversification (related to climatic and biotic vagaries, particularly in fragile ecosystems and commodity fluctuations) by expanding locally adapted or introducing novel varieties and related production systems, will contribute to improve food security and income generation for resource poor farmers and protect the environment. Small family farms will not be able to increase their total income to acceptable levels with the production of staple food crops as these are of invariably low value for the farmer or producer. To increase income the farmer needs a higher value product that can be obtained by adding value to primary or secondary products. Fruits, vegetables, herbs and spices, flavourings, natural colourants, medicinal plants and others offer opportunities for farmers to produce higher value products. Farmers would introduce new crops on their own provided the entire technological and commercial packages are introduced at the same time.

The primary constraints achieving food security in developing countries are the low yield per unit area, high population-pressure and negligible scope for expansion of the area of land for cultivation. Under these circumstances, available options will be crop intensification and diversification through the use of following instruments: (a) modern technologies, especially fertilizers, seeds and irrigation; (b) mechanisation of agricultural production; (c) development of new technologies by research; (d) farmers' participation in the planning process; (e) controlled environment agriculture; and (f) selection of high-value crops.

4.2.1 Components of Diversification

Diversification at the macro level generally means that secondary and tertiary sectors of an economy progressively become more important over time than the primary sector, in terms of their contributions to national income as well as in disposition of the work force (Shome 2009). The diversification can be of three broad categories: (a) employment diversification; (b) crop diversification; and (c) resource diversification. It has four components of growth: area, yield, pattern, and price. The growth in the crop sector can be influenced by change in area (altering the cost of production), change in the prices of inputs (altering cost of obtaining inputs); and change in technology and input intensification (altering output prices). In addition to these components, diversification or the cropping pattern mix can also influence growth by altering the allocation of resources.

4.2.2 Perspectives of Diversification of Crop Sector

Diversification in the crop sector is analysed from two analytical perspectives. One relates to the optimal crop mix on a production possibility frontier, and the other confines to mechanism for incorporating risk aversion into a farmer's decision-making process. Thus, crop specialisation may lead to highly unstable income due to variance in yield, production or price for the particular crop (World Bank 1988).

The very first dimension relates to the choice of the number of crops. When a farmer decides the number, it influences the aggregate level of spread and concentration at the farm level. This determines the overall extent of specialisation/diversification of the farmer. Here, the aspect of diversification is observed not only in terms of the number, but also in terms of the balance among the different sources (Chaplin 2000).

Second, a household with two crops would be more diversified than a household with just one crop. However, in the latter case, the diversification would be more significant between two households with two crops due to extreme difference in the share of crops to total crops output (Ersado 2003; Joshi et al. 2003). The third dimension of diversification relates to the typology of cropping pattern or crop-mix. There can be a different combination of low- or high-value crops or a mix of subsistence and commercial crops.

4.2.3 Spatial and Temporal Dimensions of Crop Diversification

In total, one may classify the spatial dimensions of crop diversification into many broad types: diversification as *multiple crops*, diversification on the basis of *level of spread or concentration*, diversification as *commercialisation* or diversification of high-value crops. There is an additional dimension (temporal) of diversification, i.e. changing the cropping pattern mix. It is important to note that the shifts among the cropping pattern can occur in any direction, i.e. from low-value crops to high-value crops or vice-versa. In terms of interpreting the significance of changing diversification (crop mix), the major debate is centered around quantifying the significance of changing crop mix in the output growth, followed by identifying the factors affecting the process of crop diversification. Minhas and Vaidyanathan (1965) were the first ones to incorporate the significance of crop mix in the output growth model. Prior to their decomposition model, only area and yield were considered the only components of growth. Minhas and Vaidyanathan (1965) provided a four-component model that captures both the static and dynamic aspect of diversification.

4.2.4 Static and Dynamic Aspects of Crop Diversification

Generally, a static aspect of diversification (initial conditions) is considered in quantifying the role of changing crop mix (diversification) on growth, while ignoring the dynamic aspect of the same that deals with the concomitant movements in the change in crop mix and technological development (Chandar 1971; Vidyasagar 1980; Joshi et al. 2006). Diversification can affect the output growth through both static and dynamic effects. If the crop pattern shifts in favour of crops, which exhibited relatively higher levels of productivity in the initial period of analysis, one gets the favourable static effect of crop shifts. The significance of such crop pattern shift can sometimes be misleading, especially if, the region also experiences decline in the total area of cultivation. The actual significance of the changing crop mix in the growth can be better captured by the dynamic aspect of changing crop mix, i.e. dynamic effect.

The dynamic effects of crop shifts capture the concomitant movements of yield and cropping pattern change. There is a possibility that when the yield of certain crops in a region goes down, farmers may leave or add area towards such crops, depending on the goals of the farmers like food security and other factors like market size (level of market development) and resource availability. We get positive dynamic effects when the shifts of cropping patterns and yields move in the same direction—either negative or positive, whereas we get the negative dynamic effects in two cases. The first case occurs when the crops, which experienced increased relative weight in the cropping pattern, experience negative growth of productivity. The second case occurs when the crops, which experienced decreased relative weight in the cropping pattern, experience higher productivity growth.

4.3 Crop Diversification and Output Growth in India

Indian economy is no longer an agrarian economy. The share of agriculture output in GDP has declined drastically from 32% in 1990–1991 to 14% in 2009–2010. The similar pattern is followed in other states. It was believed that the economic reforms in the 1990s would bring drastic change in the agricultural sector, as the trade in agricultural commodities will increase.

During the 1960s, the Indian economy was confronting the problem of food insecurity. The inadequate development of market (infrastructure) and technology were mainly responsible for low output and higher variability in the returns of both the food and non-food crops. The introduction of Green Revolution was a step towards changing the orientation of farmers in terms of the adoption of new and better technology. The technological mission on oilseeds in the late 1980s was an additional effort to improve another area of the agricultural sector. Both the efforts concentrated on technology-led development. These steps were significant both in increasing food production in India, and in increasing the growth of output, especially during 1970s and 1980s. However, the success did not sustain longer. Since mid-1990s onwards, the country started facing low growth along with near stagnation in food production (Chand 2005).

Thus, the module of development concentrated on a few food crops. Accordingly, several states showed patterns towards increased specialisation in a few crops. Such development initiative or policy came under pressure, especially after the introduction of World Trade Organization (WTO) in 1995 that demanded reduction in support measures and subsidies. At the same time, a sustained economic growth, rising per capita income and growing urbanisation caused a shift in the consumption patterns in favour of high-value crops. All these substantiated the role of diversification as a policy tool for development in the agricultural sector (Vyas 1996; Kumar and Mruthyunjaya 2002; Joshi 2005).

The question that arises is what is the role of changing crop mix in the growth of output in India and whether the changing pattern of diversification that has taken place in the last three decades, has really been growth-inductive or depressive? What are the factors that have contributed to the typology of the crop diversification in India?

4.3.1 *Data and Analytical Technique*

The study is based on secondary data obtained from different publications of Government of India, Centre for Monitoring Indian Economy, etc. Data of 30 crops is used to measure the different typologies of diversification, i.e. number of crops, level of spread or concentration and changing proportion of crop groups, where rice and wheat are characterised as one group of high-value crops and non-food grains as another. The gross value of output of these crops is obtained from the Central Statistical Organisation (CSO). Income levels are the gross value of output of these 30 crops at a constant price (1993–1994), where risk is calculated by coefficient of variability of gross value of output.

The growth rate in the crop output has been computed by the compound growth rate of the overall output in the last three decades. We have used the decomposition model provided by Minhas and Vaidyanathan (1965) to explain the components of agricultural growth in both static and dynamic ways. It is important to note that the components (area, yield, cropping pattern and interactions) are not supposed to convey the misleading impression of causation in any way, but explain only a particular form in which the growth of output took place. We obtain the interaction terms because of the multiplicative nature of the identity we started with (Chandar 1971). This serves as the dynamic aspect of agricultural growth. The observed increase in aggregate output is decomposed into four component elements, i.e. the contribution of: (i) changes in area; (ii) changes in per acre yields; (iii) changes in cropping pattern and (iv) the interaction of the two latter elements. For measuring the trend in area expansion and substitution, we have used the method by Venkataramanan and Prahladachar (1980).

Let us assume a situation of unchanged cropping pattern, where the respective areas under all crops bear the same proportion to GCA over the years. It implies that the rate of growth under individual crops must equal the rate of growth in the GCA over the same period. Such a change can be expressed in the form of a linear homogenous GCA function, where given proportionate changes in area under individual crops are related to equal proportionate change in GCA. The rates of growth in the area of individual crops, which differ from the rate of growth of GCA, therefore provide evidence of change in the cropping pattern.

The area-GCA elasticity, which can be defined either, as the ratio of the rate of growth of area under a crop to the rate of growth in GCA, or as the ratio of the area under the crop to the GCA before and after the change, can be used to measure the shift in the cropping pattern. The total change in the cropping pattern over time is the sum of the total of the substitution effect, the relative decline in area under some crops and corresponding equivalent increase in area under other substitutable crops for a given GCA and the expansion effect (effect of increase in the GCA). Statewise patterns of crop diversification, income and risk in India are given below.

Table 4.1 Number of crops produced across states in India. (Source: Centre for Monitoring Indian Economy (CMIE) and Indian Agricultural Statistics (various issues))

States	Number of crops	States	Number of crops
Jammu and Kashmir	13	Bihar	20
Punjab	14	Maharashtra	23
Haryana	15	West Bengal	23
Himachal Pradesh	18	Assam	24
Kerala	18	Orissa	24
Rajasthan	18	Tamil Nadu	24
Gujarat	19	Andhra Pradesh	25
Madhya Pradesh	19	Karnataka	27
Uttar Pradesh	19	All India	30

4.4 Results

4.4.1 Pattern of Crop Diversification in India

The intensity of diversity is reflected by the number of crops produced in a state as well as by the aggregate level of spread or concentration. The state-wise pattern of diversity on the basis of 30 crops reveal that most of the states in the northern region fall under the category of states producing less number of crops and hence are less diverse, whereas almost all the eastern and southern states are highly diverse as they produce relatively more number of crops (Table 4.1).

4.4.2 Level of Spread Across States

Interestingly, when we measure the level of spread across states in India, both the northern and eastern regions show higher levels of concentration (Table 4.2). We have categorised the states into three broad patterns based on their movement towards spread and concentration over time in their cropping patterns. The first pattern includes states/regions that have shown almost no change in their concentration index. In the second pattern, we have states, which are moving towards higher level of concentration in their cropping pattern. The third category consists of the regions that are increasing their spread among the crops.

4.4.3 Index of Concentration: State Wise

Overall, the regions that are showing stagnancy in their index of concentration or spread include Jammu and Kashmir, Kerala and Rajasthan. States from the northern and eastern regions are the ones showing increasing trend towards higher concentration, including Punjab, Haryana and Uttar Pradesh, whereas most of the southern states fall under the category of increasing trend towards spread of cropping pat-

Table 4.2 Trend towards specialisation/diversification in India. (Source: Centre for Monitoring Indian Economy (CMIE) and Indian Agricultural Statistics (various issues))

States	1970s	1980s	1990s	2000	Change in magnitude
Static spread/concentration					
Jammu and Kashmir	0.703	0.721	0.717	0.718	0.015
Kerala	0.780	0.798	0.792	0.782	0.002
Rajasthan	0.796	0.805	0.816	0.786	−0.011
Increasing concentration					
Bihar	0.517	0.468	0.450	0.427	−0.090
Haryana	0.825	0.800	0.791	0.757	−0.068
Himachal Pradesh	0.711	0.664	0.646	0.645	−0.065
Madhya Pradesh	0.855	0.836	0.823	0.817	−0.038
Orissa	0.332	0.499	0.317	0.280	−0.052
Punjab	0.749	0.660	0.634	0.612	−0.137
Uttar Pradesh	0.833	0.779	0.760	0.733	−0.100
Increasing spread					
Gujarat	0.834	0.863	0.875	0.863	0.028
Andhra Pradesh	0.801	0.802	0.798	0.838	0.036
Assam	0.425	0.489	0.458	0.460	0.034
Karnataka	0.874	0.873	0.888	0.904	0.031
Maharashtra	0.795	0.797	0.819	0.838	0.043
Tamil Nadu	0.768	0.767	0.785	0.819	0.052
West Bengal	0.385	0.434	0.417	0.476	0.090

tern. It is interesting to note that Assam, that has higher concentration index, over the period, is shifting towards more spread in the cropping pattern. However, it is not strictly the case that, states with large number of crops also have higher index of spread.

4.4.4 Pattern of Diversification of Area Under Crop Groups

The pattern of diversification on the basis of area under various crop groups shows that most of the eastern and northern regions have higher areas than two major crops i.e. rice and wheat, which is responsible for their higher index of concentration. In contrast, the southern and western regions have higher allocation of area under non-food grains (Table 4.3).

The temporal picture of changing proportionate areas under these crop groups reveals a very interesting picture. Haryana and Uttar Pradesh (so-called specialised states in India), along with Rajasthan, Gujarat and Bihar, have increased their allocation of area under both the crop groups, i.e. rice/wheat and non-food grain crops. They have replaced other inferior food crops (like cereals), in order to increase the overall allocation of non-food crops in their cropping pattern. However, in Punjab, the allocation of specialised crops (rice and wheat) has increased and the proportion of

Table 4.3 Proportionate area under high-value crops and their changing patterns. (Source: Centre for Monitoring Indian Economy (CMIE) and Indian Agricultural Statistics (various issues))

States	Proportion of area under rice and wheat (TE 2000)	Proportion of area under non-food grains (TE 2000)	Change in proportion of area under rice and wheat 1971–2001	Change in proportion of area under non-food grains 1971–2001
Andhra Pradesh	32.43	42.81	−2.81	15.38
Assam	74.58	24.65	−2.69	2.72
Bihar	74.43	9.95	5.57	0.93
Gujarat	15.18	59.19	2.33	10.94
Haryana	59.88	24.27	25.67	10.92
Himachal Pradesh	55.34	3.84	4.08	−0.69
Jammu and Kashmir	55.21	7.90	−1.73	2.62
Karnataka	15.89	41.66	−1.80	8.09
Kerala	12.62	87.34	−21.17	21.25
Madhya Pradesh	45.58	14.35	6.81	−3.78
Maharashtra	14.87	40.83	−0.67	3.58
Orissa	84.83	9.59	3.49	−1.06
Punjab	86.62	10.49	28.07	−9.39
Rajasthan	16.77	22.36	0.71	9.55
Tamil Nadu	36.52	42.59	−5.42	10.78
Uttar Pradesh	68.49	15.65	15.34	2.39
West Bengal	76.58	22.18	−7.28	9.38

TE Trinneum ending

area under other high value non-food crops have reduced. West Bengal, Tamil Nadu, Maharashtra, Jammu and Kashmir, Assam and Andhra Pradesh reduced the proportion of rice or wheat in order to increase the proportionate area under non-food crops.

4.4.5 *Income and Risk Pattern: Typologies of Diversification*

Since, the initial differences in the gross value of output across states is expected to differ according to the initial conditions and other bio-physical factors, for the purpose of comparison, we have used the changing relative ranks of the states, in terms of the level of gross value of output over the last three decades. The results show that the regions that have shown higher extent of concentration (Punjab and Haryana) are the ones, which improved in their relative ranking, in terms of gross value of output levels, of which Uttar Pradesh retained the first position (Table 4.4).

States of the eastern regions, barring Assam and West Bengal, in addition to Kerala and Tamil Nadu of the southern region (which have reduced substantial area under food crops and have been one of the major states in increasing area under non-food crops), have shown a decline in their ranking. This shows that merely increasing the proportion of area under non-food crops is not sufficient to increase

Table 4.4 Changing ranks of states in terms of gross value of output (at 1993–1994 prices). (Source: Central Statistical Organisation (various issues))

States	Rank in TE 1973	Rank in TE 1983	Rank in TE 1993	Rank in TE 2001
Bihar	7	10	11	12
Gujarat	5	4	10	10
Kerala	10	12	13	13
Madhya Pradesh	3	6	5	8
Orissa	14	14	14	15
Tamil Nadu	4	8	6	7
Assam	15	15	15	14
Haryana	13	13	12	11
Karnataka	6	7	8	5
Maharashtra	9	3	4	3
Punjab	11	5	3	4
Rajasthan	12	11	9	9
West Bengal	8	9	7	6
Andhra Pradesh	2	2	2	2
Himachal Pradesh	17	17	17	17
Jammu and Kashmir	16	16	16	16
Uttar Pradesh	1	1	1	1

TE Total estimate, Rank 1 is the higher rank and vice-versa

income. It is also essential to address the factors which help in increasing the value of output of the region.

4.4.6 *Variability in Gross Value of Output*

The variability in the gross value of output in different decades reveal that an average variability in the gross value of output has indeed declined during the decade of 1990s, as compared to the decade of 1980s (Table 4.5).

Barring Madhya Pradesh and Karnataka, all other states experienced less variability in their gross value of output during the 1990s as compared to the 1980s. This happened even for the states, which had shown a trend towards either more spread or more concentration in their cropping patterns. Also, states with less number of crops, including Jammu and Kashmir, Kerala and Himachal Pradesh, are the ones with lesser variability in the output for 1971–2001. It is important to note that the highest growth in the gross value of output during 1980s was complemented with higher variability of the same.

Table 4.5 Coefficient of variation in gross value of output across states in India. (Source: Central Statistical Organisation (Various issues))

States	1970s	1980s	1990s	1971–2001
Kerala	1.63	2.42	2.34	2.59
Jammu and Kashmir	3.29	2.93	2.31	2.66
Himachal Pradesh	2.45	4.16	2.33	3.09
Bihar	2.96	4.44	3.48	3.36
Assam	2.02	2.04	1.83	3.65
Orissa	4.25	5.29	4.52	3.73
Uttar Pradesh	3.18	2.86	2.03	4.64
Madhya Pradesh	5.02	3.41	4.82	4.65
Tamil Nadu	3.45	5.07	3.05	4.76
Andhra Pradesh	4.09	4.70	3.17	5.00
Maharashtra	7.22	4.49	3.87	5.16
Karnataka	3.22	3.48	4.01	5.28
Punjab	5.26	4.55	2.07	5.50
Gujarat	7.83	9.68	7.21	5.62
West Bengal	2.82	6.28	3.23	5.80
Haryana	6.11	5.47	2.26	6.06
Rajasthan	5.51	6.21	4.72	6.13
All India	2.73	3.44	2.22	4.56

Table 4.6 Relation between typology of diversification, income and risk pattern. (Source: Central Statistical Organisation (various issues), CMIE and Indian Agricultural Statistics (various issues))

Correlation coefficients	1970s	1980s	1990s	1971–2001
Income and risk	0.535	−0.088	−0.305	0.540
Spread and income	−0.294	0.115	0.417	−0.072
Spread and risk	−0.057	0.040	−0.063	0.121
Diversification towards high-value crops and income	−0.119	−0.072	−0.034	−0.366
Diversification towards high-value crops and risk	0.142	−0.312	0.151	−0.243
Diversity and income	–	–	–	0.207
Diversity and risk	–	–	–	0.121

Note: Indicators for income=gross value of output, risk=viability in gross value of output, spread=Herfindahl index (1—more spread), Diversification towards high-value crops=proportionate area under non-food crops, diversity=number of crops

4.4.7 Heterogeneity in Pattern/Typology of Diversification

We observe great heterogeneity in terms of the pattern or typology of diversification across states. For instance, some states are highly diverse in their cropping patterns and also have lesser proportion of non-food crops in their cropping patterns. We examined the relationship between different patterns of diversification with income and risk by correlation coefficient (Table 4.6).

Overall, higher changes in the income are complemented with increased variability in returns, as the correlation coefficient between gross value of output and its variability turns out to be 0.540. Interestingly, there is an inverse and positive relationship of increasing spread of the cropping pattern with income and risk, respectively.

During the period of liberalization (1990s), we observed a different picture. The regions have gained by increasing spread in the cropping pattern through high income with less risk. This shows that specialisation has indeed proved a deterrent for growth, especially after liberalisation in India. We do not find positive link of high proportion of area under non-food grains with income levels, as the correlation coefficient is negative in all the three decades, including the last decade of liberalisation. However, the regions showing higher proportion of non-food crops have faced lesser risk during 1990s. Overall, it is found that the regions that are showing trends of shifting towards non-specialised crops are getting benefitted. Diversity in the cropping pattern is noted to be positively correlated with both income and risk, which shows that by merely increasing diversity, the overall risk may not be reduced. It also requires other typologies of diversification, which are also critical.

4.4.8 Relationship Between Crop-Mix and Output

The trends in growth of output across states for the past two decades reveal that very few states have been able to experience higher output growth in both the decades (Table 4.7). West Bengal, Himachal Pradesh and Maharashtra were able to achieve growth rate above India's average in both the decades.

Some of the states like Bihar, Jammu and Kashmir and Karnataka were able to improve their growth during 1990s as compared to 1980s. Interestingly, the major agricultural states like Punjab, Haryana and Uttar Pradesh, which were able to gain from the Green Revolution, showed decline in the growth rate during 1990s, whereas some states like Assam, Kerala and Andhra Pradesh continue to exhibit poor growth.

The trends indicate that regions which were able to show higher levels of diversification towards high value food grains and were also major contributors to overall growth, have been experiencing plummet in the growth rate, whereas, some backward regions continue to exhibit poor growth in this sector and have not been able to benefit either from Green Revolution or liberalisation regime. At the same time, we have a few regions, which, though, did not gain from Green Revolution, are now picking up growth in the recent past. Such a distinct pattern of growth across region and time increases the importance to quantify the major components of growth in the regions across different periods.

Table 4.7 Trend in the output growth rate across states in India. (Source: Central Statistical Organisation (various issues))

States	Growth rates (%)
High agriculture growth states (1980–1990)	
West Bengal	5.76
Punjab	5.16
Haryana	4.44
Rajasthan	4.31
Maharashtra	3.85
Madhya Pradesh	3.65
Tamil Nadu	3.49
Himachal Pradesh	3.46
Low agriculture growth states (1980–1990)	
Uttar Pradesh	2.93
Kerala	2.05
Bihar	2.60
Karnataka	2.50
Assam	2.39
Andhra Pradesh	2.30
Orissa	1.46
Jammu and Kashmir	1.20
Gujarat	−0.81
High agriculture growth states (1991–2001)	
Bihar	4.27
West Bengal	3.85
Himachal Pradesh	3.69
Maharashtra	3.65
Jammu and Kashmir	3.45
Karnataka	3.26
Low agriculture growth states (1991–2001)	
Assam	2.75
Rajasthan	2.70
Haryana	2.67
Kerala	2.62
Andhra Pradesh	2.34
Uttar Pradesh	2.21
Tamil Nadu	2.11
Gujarat	1.81
Punjab	1.27
Orissa	0.46
Madhya Pradesh	−0.03

4.4.9 Growth of Crop Output in India and Across Regions

Changes in cropping patterns help to accommodate risks, originating from various factors, as well as optimise income either by reducing cost or by moving towards high-value crops. In an aggregate picture, any change in cropping pattern also

Table 4.8 Contribution of different components of growth to the crop sector by various regions in India, 1971–2001 (%). Source: Author's computation based on the data obtained from Central Statistical Organisation (various issues), CMIE and Indian Agricultural Statistics (various issues)

Region	Period	Area effect	Inter crop shift effect (static)	Aggregate yield effect	Inter crop shift effect (dynamic)
Northern zone	1970s	28.45	8.65	54.34	8.56
	1980s	10.35	11.54	72.54	5.57
	1900s	32.54	18.54	43.25	5.67
	1971–2001	23.45	12.65	60.12	3.78
Eastern zone	1970s	43.65	30.15	30.78	−4.58
	1980s	24.56	8.36	73.80	−6.72
	1900s	45.25	7.10	45.35	2.30
	1971–2001	23.15	10.05	64.00	2.80
Western zone	1970s	18.33	23.85	58.64	−0.82
	1980s	15.48	21.07	61.54	1.91
	1900s	−25.85	87.65	38.45	−0.25
	1971–2001	19.95	48.25	30.54	1.26
Southern zone	1970s	−20.85	30.04	87.05	3.76
	1980s	4.34	28.65	62.15	4.86
	1900s	−21.00	86.54	35.26	−0.80
	1971–2001	−4.15	48.56	58.45	−2.86
All India	1970s	19.78	14.66	62.12	3.44
	1980s	14.33	19.39	69.17	−2.89
	1900s	11.29	58.65	27.66	2.40
	1971–2001	17.33	24.66	61.52	−3.51

causes a shift in the aggregate production function of the sector, and the process of diversification is responsible for this. In order to understand the full effect of diversification, it is necessary to approach through decomposition of the rates of growth that not only provide the contribution of the components of growth but also help to indicate their relative positions.

Contribution of area, yield and change in the cropping patterns on the gross crop income across regions are computed for four broad regions. The results of the sources of growth suggest that diversification (crop-pattern mix) has indeed become an important source of growth in the post-liberalisation period, and technological development has made the largest contribution to the output growth in the previous two decades (Table 4.8).

The contribution of crop area as a source of growth is showing a declining trend in its contribution to aggregate output over the period of time. However, the region-wise picture revealed that area expansion is still the major contributor to the growth of the eastern region of the country. Expansion of area through extensification or intensification has been the major source of growth of output and crop pattern change, which is a cause for concern as it is less viable and unsustainable in the long run. Crop pattern shift has minimal role in the growth of output as this region continues to diversify towards the highly specialised crop, i.e. rice.

Table 4.9 Area expansion and substitution effects in India, 1971–2001. (Source: Author's computation based on the data obtained from Central Statistical Organisation (various issues), CMIE and Indian Agricultural Statistics (various issues))

Year	India (000 ha)	In percentage to total			
		North	East	West	South
Area expansion					
1971–2001	21299	34.63	9.16	42.66	13.45
1970s	5817	55.82	13.43	27.36	4.05
1980s	6733	23.32	20.70	31.54	25.21
1990s	5840	34.22	16.24	30.32	19.32
Area substitution					
1971–2001	9312	39.17	5.97	26.43	28.35
1970s	1928	59.05	5.06	6.04	30.21
1980s	3338	20.73	7.53	29.52	43.20
1990s	4988	25.99	2.45	41.24	30.88
Net area change					
1971–2001	11987	35.88	10.85	57.74	−4.52
1970s	3888	50.88	15.60	55.64	−21.66
1980s	3394	16.04	35.64	37.58	10.58
1990s	851	127.14	97.02	−57.92	−65.93

Both, the southern and western regions, are experiencing declining contribution of area in their output growth and their contribution became negative during the 1990s, as the total area under cultivation declined. Decline in the area under cultivation was compensated through increased diversification towards high-value crops, rather than by productivity growth, which resulted in high diversification effect.

In the northern zone, though the role of crop pattern shift is increasing over time, the contribution of the same is still low. However, this region has recorded high growth in the shift of cropping patterns. The reason for getting low diversification effect is that change in the cropping patterns has resulted in the increased specialisation towards these crops. At the same time, the yield of these crops increased manyfold, which resulted in the increased role of yield.

Briefly, technological development continues to be the major component of growth in the northern region, but its effect is declining. This is primarily due to decline in the yield of the specialised crops. For eastern region, area expansion is the major source of growth in the absence of any significant technological development and crop shifts. For both the southern and western regions, diversification has become an important component for growth, primarily, due to declining area for cultivation and low technological development.

Crop diversification has become an important component of growth, especially in the southern and western region of the country. Since, role of changing allocation of areas in the growth of output (diversification) can also be influenced by changing growth of area; we measured both, the area expansion and the area substitution pattern, across these four broad regions in India (Table 4.9).

Indian agriculture during the last three decades experienced continuously increasing the substitution effect. The 1980s saw increase in area expansion effect, but the same declined during the 1990s. This indicates that the avenues of increasing growth through area increase are rather dim and there is a need to look for other factors for stepping up growth.

Northern and western regions have demonstrated a strong area expansion, whereas the eastern region has shown poor performance in terms of both area expansion as well as crop substitution. Thus, crop substitution is becoming more important in the western and southern states of the country, which have also shown negative net area change during the recent decade (1990s), and has resulted in getting an increased role of crop diversification in the growth of output in these regions.

High importance of any of the individual components provide the static effect of the components, whereas the dynamic effect, i.e. interaction effect, helps to reveal whether over the period of time, the shifts in the cropping patterns takes place towards the crops, which also experience higher technological development. The results reveal that northern region is the only one, which through all the decades has been able to improve the cropping patterns towards the crops, and has experienced better technological growth. Whereas, all other regions have shown mixed trend over the period, the eastern region has been able to achieve positive growth in the productivity of the crops, towards which it had diversified during 1990s.

The western and southern regions have experienced high and negative interaction effect during 1990s, raising the concerns of the lack of technological development in the enterprises. Though, there has been increased growth in the cropping pattern shift towards high-value crops during 1990s, the growth rate of those yields declined dramatically during the same period, which has resulted in the growth-depressing effect of diversification in the western and southern regions.

This suggests that diversification towards high value per se is not sufficient for increasing growth. These crops should remain to be remunerative over the period of time, through proper technological and market development, otherwise the gains from diversification will be meagre.

The typology of state-wise picture of crop diversification shows that the cumulative number of states that have shown an increased importance of crop like rice, wheat, sugarcane, and horticultural crops, in their cropping patterns, has increased significantly during 1990s as compared to 1980s (Table 4.10).

The number of states having positive dynamic impact across these crop groups declined during 1990s, as the cumulative frequency of states showing positive dynamic effect declined from 76 to 65. This is the reason, why, despite increased shift in cropping patterns towards high-value crops during 1990s, we experienced poor agricultural growth in India.

For northern and eastern regions, emphasis on increasing the crop diversification towards high-value crop includes many of the non-food grains. This was due to increased vulnerability of these regions to external shocks and high level of specialisation in rice and wheat. For southern and western regions, intervention is required to improve the technology of many of the high-value crops, which are becoming important for the growth of the region.

For this purpose, we use correlation coefficient, which is measured separately for the absolute area change in rice and wheat as one group, and for non-food grains as

Table 4.10 Typology of dynamic shifts across states in India, 1971–2001. (Source: Author's computation based on the data obtained from Central Statistical Organisation (various issues), CMIE and Indian Agricultural Statistics (various issues))

Particulars	Period	Rice	Wheat	Coarse cereals	Pulses	Oilseeds	Sugar cane	Cotton	Other crops	F&V	Total
Number of states with positive level of diversification (increase in area)	1971–2001	10	10	1	9	13	7	11	13	13	87
	1970s	11	13	3	10	8	10	13	14	13	95
	1980s	10	7	3	10	15	12	14	9	11	91
	1990s	13	12	4	7	5	14	13	12	15	95
Number of states having positive diversification and positive dynamic effect	1971–2001	10	9	1	9	8	7	10	13	12	79
	1970s	10	11	3	9	6	7	12	8	9	75
	1980s	9	7	3	6	13	9	12	7	10	76
	1990s	8	10	4	6	1	10	8	8	10	65

Note: Figures are in numbers; *F&V:* Fruits & Vegetables

Table 4.11 Factors influencing nature and direction of diversification in India, 1971–2001. (Source: Author's computation based on the data obtained from Central Statistical Organisation (various issues), CMIE and Indian Agricultural Statistics (various issues), Fertiliser Statistics, Statistical Abstract, India, Ghosh and De 2005 and Indian Labour Statistics (various issues))

Correlation coefficient	Rice and wheat		Non-food grains		Total
	Static (area)	Dynamic	Static (area)	Dynamic	Dynamic
Fertiliser kg/ha	0.4092	0.4926	−0.1578	0.2507	0.4070
Gross irrigated area	0.2901	0.1725	0.4427	0.5323	0.4820
Coefficient of variation of rainfall	0.3335	0.3506	−0.0446	−0.1284	−0.0078
Total workers (agricultural labourers and cultivators)	−0.1269	−0.0005	0.4711	0.2868	0.2155
Rural literacy rate	−0.0002	−0.1237	0.4450	−0.1247	−0.0332
Number of regulated markets	0.2858	0.4927	0.2262	0.3128	0.2010
Roads (km)	−0.1033	−0.2659	0.3193	0.1250	0.4180
Tractor (number)	0.6092	0.6085	0.1195	−0.0728	0.1171

another, with several factors like input development, market development and socio-economic development. Secondly, correlation coefficient is measured separately for concomitant shift of area and yield of respective crop groups (dynamic effects) as well as for all crops as one group, with the above stated factors (Table 4.11).

The coefficient correlation between area growth and input development reveals that both, the use of fertilizer and irrigation development, are important for the area increase of rice and wheat. For non-food grain crops irrigation is important because many of these crops are highly water-intensive and require development of irrigation. The regions that have experienced higher variability in rainfall over the last three decades have failed to diversify towards non-food grain crops.

The variability in climate has influenced the process of crop diversification and hence, the growth of output in India. The availability of number of workers across the states has significant influence on the diversification towards non-food grain crops, but not on the diversification towards rice and wheat. This is expected, as the non-food grain crops are highly labour-intensive, as compared to rice and wheat.

The improvement in rural literacy is also found to be important for non-food grain crops, but not for food crops like rice and wheat. Development of markets also exerted an influence on the diversification towards either of the crop groups in India.

4.5 Summary and Conclusions

There is no doubt that significant progress has been made during the past few years in crop diversification. The fundamental principal of all successful diversification programs is that they are driven by market demand. There is no point in diversifying into a crop which has limited market potential. Crop diversification can be approached in two complementary and interactive ways (horizontal and vertical). Vertical diversification is complementary to horizontal diversification, and opportunities should be explored for product diversification and value addition to achieve highest economic returns.

In order to achieve the benefit the process of diversification should be changed from very simple forms of crop rotations to intensive systems such as relay cropping and intercropping or specialisation by diversifying into various crops, where the output and processing, etc., could be different. This process could be similar at the farm and national levels. There is a need and scope to further promote crop diversification on scientific lines and realize its untapped potential.

Diversification is one of the vital policy tools, especially in the present era of globalisation, due to both supply and demand side-factors. However, there exists a hazy picture regarding the actual dimension of diversification that is critical in achieving the goal of high growth of output through high income and less risk in the agricultural sector. There is a link between different dimensions of diversification and the growth of output in India, in the last three decades. The results show that there is a mixed picture regarding the typology of diversification within the states. An inverse and positive relationship of increasing spread of the cropping pattern with income and risk respectively is found.

Interestingly, positive link of high proportion of area under non-food grains and income levels is not found, as the correlation coefficient is negative in all the three decades, including the last decade of liberalisation. However, the regions showing higher proportion of non-food crops have faced less risk during the 1990s. Diversity in the cropping pattern (number of crops) is positively correlated with both income and risk, which shows that by merely increasing diversity, the overall risk may not be reduced. Rather, it also requires other typologies of diversification, which are crucial.

The relative position of diversification (change in crop mix as against area and productivity) is a component of growth of output in India, especially in the post-liberalisation period, whereas technological development was mainly responsible for the output growth during the previous two decades, i.e. the 1970s and the 1980s. Though crop diversification especially in the southern and western regions has negative net aggregate area, it has enhanced the importance of crop substitution and hence, increased the importance of crop pattern mix as static variable.

The results of the interaction terms revealed that the northern region is the only one, which, through all the decades has been able to improve the crop pattern towards the crops, and has experienced better technological growth. Rest of the regions have shown mixed trends during the study period. The results prove that the diversification of crops remain to be remunerative during the study period.

For northern and eastern regions, emphasis is laid on increasing the crop diversification towards high-value crops that include many non-food grains. It is because of the increased vulnerability of these regions to external shocks, due to higher level of specialisation in rice and wheat.

For southern and western regions, intervention is required to improve the technology of many of the high-value crops, which are becoming important for the growth of the region. To improve the nature and direction of diversification in India, both fertilizer and irrigation development are important for rice and wheat crops.

Irrigation is important for non-food grain crops, as many of these crops are highly water-intensive and require assured irrigation. High labour intensity of these high-value crops also demands more availability of workers for improving the diversification towards high-value crops. Finally, development of markets is vital in order to drive diversification towards either of the crop group (rice, wheat or non-food grains) and to improve the aggregate dynamic effect in India.

Crop diversification is one of the best options to increase farm income leading to food, nutrition and ecological security as well as poverty alleviation in the Asia–Pacific region. Therefore, greater attention should be paid to crop diversification by the governments of the region.

Governments can take several steps to reduce risks and improve marketing facilities through improved roads and communications, construction of wholesale markets, etc. Access by farmers, private traders and exporters to credit, also needs to be improved. Dynamic policies on balance between food self-sufficiency and food self-reliance will be needed.

Efforts should be made by different countries to identify high specialty crops, new crops, off-season varieties and production systems, and novel varieties of crops with comparative advantage, mainly fruits and vegetables, to open up new opportunities for farmers.

The promotion of multipurpose species would also be useful for diversification of agro-processing on small-scale at local and national level for productivity enhancement and expanded employment opportunities. Again, there is a need for improved seed and other planting materials for effective crop diversification. One major concern is the high post-harvest loss of crop produce, particularly in horticultural crops, in most countries.

The governments of all countries should take initiatives to minimise such loss. The private sector can play a major role in the development of modern agro-enterprises to infuse capital and technology into diversified cropping systems for effective commercialisation for long-term sustainability.

Since crop diversification is an important element of poverty alleviation, income generation, equity and natural resource conservation, hence to enhance this, a well-designed mechanism has to be developed through the participation of international organisations and local governments to strengthen the initiative undertaken by this region. There is also a need for development of an information database on crop diversification for policy makers, farmers, consumers and other stakeholders.

References

Central Statistical Organisation (CSO) (various issues) Government of India, National Accounts Statistics, New Delhi, India

Centre for Monitoring Indian Economy (CMIE). Economic intelligence service. http://www.cmie.com (Various years). Accessed 12 June 2011

Chand R (2005) Exploring possibilities of achieving four percent growth rate in Indian agriculture. http://ncap.res.in/upload_files/others/oth_7.pdf. Accessed 28 August 2011

Chandar P (1971) District-wise measurement of decomposition of the growth of agricultural output in the Punjab during post-independence period. Indian Econ Rev 6(1):49–63

Chaplin H (2000) Agricultural diversification: a review of methodological approaches and empirical evidence. www.agp.uni-bonn.de/agpo/rsrch/dara/Farm/wyewp2.doc. Accessed 28 August 2011

Ersado L (2003) Income diversification in Zimbabwe: welfare implications from urban and rural areas, FCND Discussion Paper No. 152, IFPRI, Washington DC

Fertilizer Statistics (various issues) Government of India, Ministry of Agriculture, New Delhi, India

Ghosh B, De P (2005) India infrastructure database (vols I and II). Bookwell Publishers, New Delhi

Indian Agricultural Statistics (various issues) Government of India, Ministry of Agriculture, New Delhi, India

Indian Labour Statistics (various issues) Government of India, Ministry of Labour, New Delhi, India

Joshi PK (2005) Crop diversification in India: nature, pattern and drivers. National Centre for Agricultural Economics and Policy Research, New Delhi. http://www.adb.org/Documents/Reports/Consultant/TAR-IND-066/Agriculture/joshi.pdf. Accessed 28 August 2011

Joshi PK, Gulati A, Birthal PS, Tewari L (2003) Agriculture diversification in south Asia: patterns, determinants, and policy implications, MSSD Discussion Paper No. 57, IFPRI, Washington, DC

Joshi PK, Birthal PS, Minot N (2006) Sources of agricultural growth in India: Role of diversification towards high value crops, MTID Discussion Paper No. 98, IFPRI, Washington, DC

Kumar P, Mruthyunjaya (2002) Long-term changes in food basket in India. J Agric Econ 6(2, April):17. (Paper presented in an international workshop on Agricultural Diversification in South Asia, November 21–23, jointly organized by MOA, NCAP, IFPRI, Paro, Bhutan, ICFAI University)

Minhas BS, Vaidhyanathan A (1965) Growth of crop output in India, 1951–54 to 1958–61: analysis by component elements. J Indian Soc Agric Stat 28(2):230–252

Statistical Abstract of India (various issues) Government of India, Central Statistical Organisation, Ministry of Statistics and Programme Implementation, New Delhi, India.

Shome S (2009) An analysis of crop diversification: experience in the Asia–Pacific region. J Agric Econ 6(1):7–30 (January, ICFAI University)

Venkataramanan LS, Prahladachar M (1980) Growth rates and cropping pattern changes in agriculture in six states: 1950 to 1975. Indian J Agric Econ 35(2):71–84

Vidyasagar (1980) Decomposition of growth trends and certain related issues. Indian J Agric Econ 35(2):42–59

Vyas VS (1996) Diversification in agricullture: concept, rationale and approaches. Indian J Agric Econ 51(4):636–643

World Bank (1988) Diversification in rural Asia, Working Paper Series No. 98, Agriculture and Rural Development Department, The World Bank, Washington, DC, USA

Chapter 5
Trend and Pattern of Crop Diversification in Odisha

R. K. Panda

5.1 Introduction

Odisha is primarily an agricultural state. Over the years, agriculture's contribution to the state's GSDP has declined sharply from 70% in 1950s to 18.44% at present. Yet this sector continues to be the dominant sector in the state economy as it provides livelihood to more than 60% of the state's total workforce. However, agriculture in the state is characterized by the predominance of small holdings. The marginal and small farmers (within 2 ha of holding) constitute more than 83% of the agricultural households and cultivate around 55% of area. Due to diverse agro-climatic conditions and frequent occurrence of natural calamities, a large number of agricultural items are produced with low productivity. Among the crops grown, paddy remains the principal crop covering nearly three-fourth of cultivated area. In recent years, however, with the state economy showing signs of higher growth and faster reduction in poverty and different policy initiatives taken by the state government to increase the cultivation of high-value crops (HVCs), diversification of agriculture is underway. Against this backdrop, the present study primarily aims at measuring the trend and pattern of crop diversification at the state and regional levels with the help of recent data. In view of the predominance of small holdings in the agricultural sector in the state the study also examines the association of small holders in diversifying crops. By way of discussion, the study outlines the recent policy initiatives taken up by the state government towards crop diversification with a conclusion and policy suggestions.

R. K. Panda (✉)
Nabakrushna Coudhury Centre for Development Studies, Bhubaneswar, Odisha, India
e-mail: rkpanda2@rediffmail.com

M. Ghosh et al. (eds.), *Diversification of Agriculture in Eastern India*,
India Studies in Business and Economics, DOI 10.1007/978-81-322-1997-2_5

5.2 Data and Methodology

The study is based on secondary data obtained from different publications of the Government of Orissa. Macrodata at the all-Odisha and regional levels are studied. For the regional analysis the state is divided into four regions taking into account the agro-climatic diversities—the Coastal Plain, the Central Table Land, the Northern Plateau and the Eastern Ghats. From amongst the regions the Coastal Plain, stretching over 11 districts, is a fertile tract well suited for intensive cultivation. The soil of this region is mostly alluvium. The rainfall is more evenly distributed in this tract and climatically it is more suitable for increasing agricultural production. The Central Table Land, consisting of eight interior districts and developed out of erosional plains and river valleys, is also a fertile tract. The soil of this region is rich in potassium and magnesium and thus suitable for growing pulses and wheat. The Northern Plateau covers three districts; it is richly endowed with mineral resources. Much of this Plateau is hilly, heavily forested and not suitable for agriculture. The Eastern Ghats, stretching over eight districts, is an uneven tract with high mountains and valleys. The topography of this region is very much undulated and the soil is lateritic in character, having formed out of the decomposition of rocks. The soil is poor in plant nutrients and hence not of much help for intensive cultivation. In measuring the trend and pattern of crop diversification, particularly changes in acreages under different crops/crop-groups are studied at the state and four regional levels. Annual compound growth rate is worked out to measure the trend analysis. Association of small holders with the area devoted to two HVCs such as fruits and vegetables in different regions is worked out to assess the extent of their participation in the crop diversification process in the state. Simple regression analysis is worked out taking the district-wise data on the area under fruit and vegetable cultivation as the dependent variable (Y) and the area cultivated by the small holders in different districts as independent variable (X) separately for each region. Since the Northern Plateau consists of three districts and has many similarities with the Eastern Ghats with respect to soil and topography, for regression analysis we have clubbed them together. The linear regression equation is fitted to the data covering the period from 2001–02 to 2009–10.

5.3 Results and Discussion

5.3.1 Changes in Cropping Pattern at the State Level

Table 5.1 presents data on changes in the cropping pattern at the state level in the selected years 2001–2002 and 2009–2010. Paddy is the staple cereal crop in the state and is cultivated widely irrespective of disparity in agro-climatic conditions across regions. However, as data reveal there has been a decline in percentage share of area under paddy from 51.14 to 48.10 in the selected years. Other food crops like

Table 5.1 Changes in cropping pattern in Odisha between 2001–2002 and 2009–2010 (area in percentage)

Crops	2001–2002	2009–2010
Paddy	51.14	48.1
Other cereals	4.99	5.10
Pulses	19.82	23.05
Oilseeds	9.65	8.79
Fibres	1.30	0.97
Spices	1.50	1.63
Vegetables	7.10	7.65
Sugarcane	0.34	0.41
Fruits	4.09	4.26
Tobacco	0.06	0.04
Total	100.00	100.00

Table 5.2 Annual compound growth rate in gross cropped area under different crops in Odisha (in percentage)

Sl. no.	Crops	Year (Area in 'thousands of hectares)	Annual compound growth rate	
	2001–2002	2009–2010		
1.	Paddy	4499.78	4365.08	*−0.05*
2.	All other cereals	439.30	463.12	*1.20*
3.	Pulses	1743.60	2092.02	*4.31*
4.	Oilseeds	849.38	797.34	*1.48*
5.	Fibres	114.66	87.82	*−0.04*
6.	Spices	132.07	147.80	*0.94*
7.	Vegetables	624.89	694.19	*1.00*
8.	Sugarcane	29.66	36.86	*4.84*
9.	Fruits	359.59	386.71	*0.06*
10.	Tobacco	5.68	3.30	*−4.99*
11.	All crops	8798.61	9074.24	*1.12*

minor cereals and pulses show an increase in percentage area between the study years. Among the cash crops while we notice a decline in percentage area in case of oilseeds, fibres and tobacco, for spices and sugarcane the area share has gone up in the selected 2 years. Vegetables and fruits exhibit a rise in their share in the gross cropped area during the period. Thus, changes in cropping pattern as noticed reveal a slow shift from cultivation of cereals to non-cereal crops in the state.

5.3.2 *Area Growth Rate in Different Crops at the State Level*

Annual compound growth rate in area under different individual crops/crop-groups is worked out for the period from 2001–2002 to 2009–2010. Observation reveals a mixed trend (Table 5.2). Among the crops, paddy, fibres and tobacco exhibit a

Table 5.3 Changes in acreages under different crops across regions of Odisha between 2001–2002 and 2009–2010 (area in percentage)

Crops	Regions							
	Coastal Plains		Central Table Land		Northern Plateau		Eastern Ghats	
	2001–2002	2009–2010	2001–2002	2009–2010	2001–2002	2009–2010	2001–2002	2009–2010
Paddy	51.36	49.61	52.72	50.45	62.02	56.14	43.06	39.48
All other cereals	3.13	3.11	1.36	1.25	4.68	4.50	11.17	11.70
Pulses	24.89	26.31	19.61	23.91	12.06	14.91	17.55	22.32
Oilseeds	6.90	5.87	12.29	11.10	7.58	8.04	12.12	11.01
Fibres	0.78	0.42	1.23	1.23	0.99	0.76	2.27	1.57
Spices	1.15	1.52	1.72	1.58	1.42	1.43	1.82	1.93
Vegetables	7.09	8.48	7.07	6.28	7.73	9.92	6.78	6.55
Sugarcane	0.45	0.53	0.27	0.19	0.04	0.02	0.42	0.65
Fruits	4.24	4.15	3.72	3.99	3.49	4.30	4.57	4.65
Tobacco	0.00	0.00	0.01	0.00	0.00	0.00	0.24	0.14
All crops	100.00	100.00	100.00	100.00	100.00	100.00	100.00	100.00

negative growth rate in the area while that for minor cereals, pulses, oilseeds, spices, fruits and vegetables and sugarcane a positive growth rate is noticed. The inference is that in the cultivation of different crops there has been a shift in area growth rate from cereals and particularly paddy to non-cereal crops in the state.

5.3.3 Regional Disparity in Crop Diversification

Changes in cropping pattern along with annual compound growth rate to measure growth in area under individual crops at the four regional levels during the selected years are presented in Tables 5.3 and 5.4, respectively. Data in Table 5.3 show that in the case of the Coastal Plain the percentage share of area under paddy, minor cereals, oilseeds, fibres and fruits has fallen while that under pulses, spices, vegetables and sugarcane has gone up during the selected years. For the Central Table Land, we notice a decline in the share of all crops excepting pulses. In the Northern Plateau, out of ten crops, the decline percentage share in area is noticed in five crops—paddy, minor cereals, fibres and sugarcane—with no changes in area under tobacco. For the Eastern Ghats share in crop area under paddy, oilseeds, fibres, vegetables and tobacco has fallen while that under crops such as minor cereals, pulses, spices, sugarcane and fruits has increased. Overall data analysis reveals that in all four regions there has been decline in the percentage area under paddy and a rise in the area under pulses. Besides, changes are noticed towards crops such as vegetables, fruits and cash crops like sugarcane and spices in all the three regions excepting the Central Table Land.

Annual compound growth rates in area under individual crops are worked out and presented in Table 5.4. In the case of the Coastal Plain, we notice positive

Table 5.4 Annual compound growth rate in gross cropped area under different crops across regions between 2001–2002 and 2009–2010 (percentage)

Regions	Coastal Plains	Central Table Land	Northern Plateau	Eastern Ghats
Paddy	0.41	−0.21	−0.78	−0.03
Other cereals	0.93	0.04	0.39	1.62
Pulses	3.15	4.34	3.12	6.97
Oilseeds	0.33	1.8	1.45	2.27
Fibres	−4.8	3.44	−2.27	0.84
Spices	1.55	0.56	0.63	1.03
Vegetables	1.52	0.09	1.84	0.5
Sugarcane	3.58	−2.82	−6.51	10.09
Fruits	−1.06	0.93	2.2	−0.19
Tobacco	NA	−18.34	NA	−4
All crops	1.06	1.07	0.31	1.75

growth rates in the area for all crops excepting fibres and fruits. Sugarcane and pulses exhibit appreciable growth rates of more than 3 % while the area growth in the case of spices and vegetables show more than 1 % growth rate. Area growth in paddy and oilseeds reveals a positive but slow growth—less than 1 %. In the Central Table Land, a positive and high growth rate in the area under pulses and fibres are observed. Paddy, sugarcane and tobacco show a negative growth in area in this region. Crops like oilseeds show more than 1 % growth in area while that of fruits, vegetables and minor cereals show positive but marginal growth in area. In the case of the Northern Plateau, the area under pulses followed by fruits and vegetables show positive and high growth rates. However, for crops like sugarcane, fibres and paddy, a negative growth rate is observed. Other crops exhibited positive but marginal growth in acreages in the region. In the Eastern Ghats pulses and sugarcane followed by oilseeds show positive and high growth in area. However, crops like paddy, fruits and tobacco show negative growth in area. In the case of other crops, the area growth rate remains positive but moderate. Thus, from the above analysis it can be inferred that there has been an indication of changes in the area from traditional cereal crops to commercial crops and HVCs in all four regions, though as we notice area growth rate in individual crop varies across regions and it might be due to disparity in agro-climatic factors.

5.3.4 Participation of Small Holders in Diversification

There is often debate among researchers on the ability of small holder-dominated subsistence farm economy to diversify into HVCs. Scholars like Longhurst (1988) opine that in laggard regions, the small farmers are at a great disadvantage to take up diversification and this happens due to a large number of causal factors including institutional, technological, etc. On the contrary, the study of Joshi et al. (2006) of the Indian agricultural development scenario reveals that small farmers do have a

Table 5.5 Proportion of marginal and small holdings and area operated by size class of holdings

Regions	1995–1996		2000–2001	
	Percentage of holding	Percentage of area cultivated	Percentage of holding	Percentage of area cultivated
Coastal Plains	86.18	57.64	86.91	62.57
Central Table Land	79.02	45.61	79.44	46.66
Northern Plateau	82.58	52.08	84.45	55.3
Eastern Ghats	76.35	43.28	78.16	45.91
Odisha	81.98	50.27	83.06	53.12

tendency towards gradual diversification, towards HVCs particularly towards vegetables, as it is very much labour intensive. As stated earlier, Odisha largely has subsistence agriculture with a high concentration of small holders (below 2 ha of holding). Data in Table 5.5 show that small and marginal farmers constituting about 83 % of holdings cultivate more than 53 % of cultivated area in the state (Economic Survey 2010–2011, Government of Odisha). Moreover as it is noticed during the period from 1995–1996 to 2000–2001 the number and area of small holders have increased in the state. Across regions the highest concentration of small holdings in number as well as area is observed in the Coastal Plain followed by the Northern Plateau, the Central Table Land and the Eastern Ghats. The concentration in number varies between the lowest 78.16 % in the Eastern Ghats to the highest 86.91 % in the Coastal Plain. In terms of area cultivated the Eastern Ghats shows the lowest concentration (45.91 %) while the Coastal Plain has the highest concentration (62.57 %) of small holders, with the other two regions lying in between.

For studying the association of small holders in the crop diversification we have taken here two crops such as fruits and vegetables. Simple regression analysis is worked out with the area under fruits and vegetables as the dependent variable (Y) and area cultivated by the small holders as the independent variable (X) for each region during the period from 2001–2002 to 2009–2010.

Regression coefficients of the independent variable estimated through simple regression analysis along with *t*-value and R^2 for area under fruits and vegetables for all the three regions are presented in Tables 5.6 and 5.7. Data analysis reveals that in the case of area under fruits the regression coefficient associated with independent variable is found to be statistically significant in Central Table Land, and Northern Plateau and Eastern Ghats (taken together). For the Coastal Plain the independent variable, explaining the dependent variable, is found statistically insignificant implying that the small farmers in this region do not influence the area of cultivation under fruits. With regard to the area under vegetables, the regression coefficient associated with the independent variable is found statistically significant in the case of the Coastal Plain and Northern Plateau and Eastern Ghats (taken together). For the Central Table Land the independent variable explaining the dependent variable is found statistically insignificant, indicating that in vegetable cultivation the small farmers do not play a significant role in this region.

Table 5.6 Estimated values of regression coefficients and other related statistics for the area under fruits in three different regions

Sl. no.	Variables	Coastal Plain		Central Table Land		Northern Plateau and Eastern Ghats	
		Regression coefficient	*t*-value	Regression coefficient	*t*-value	Regression coefficient	*t*-value
1.	Constant (a)	−0.4461		−2.0715		−2.2118	
2.	Area of small holders (X)	0.3012	0.7835	0.6364	2.3244[a]	0.6767	2.8277[b]
3.	R^2	0.0638		0.4738		0.4704	
4.	N	11		8		11	

Coefficients without asterisk are not statistically significant
[a] Coefficient is statistically significant at the level of 0.10
[b] Coefficient is statistically significant at the level of 0.05

Table 5.7 Estimated values of regression coefficients and other related statistics for area under vegetables in three different regions

Sl. no.	Variables	Coastal Plain		Central Table Land		Northern Plateau and Eastern Ghats	
		Regression coefficient	*t*-value	Regression coefficient	*t*-value	Regression coefficient	*t*-value
1.	Constant (a)	−1.6307		−1.9930		−2.1990	
2.	Area of small holders (X)	0.6077	2.8631[a]	0.6572	1.8040[a]	0.7195	2.1234[a]
3.	R^2	0.4766		0.3516		0.5824	
4.	N	11		8		11	

Coefficients without asterisk are not statistically significant
[a] Statistically significant at the 5 % level

5.3.5 *State Government Policy Towards Diversification of Agriculture*

In recent years, the state government has taken a number of policy initiatives to diversify agriculture and particularly divert land from paddy to non-paddy and commercial crops. In 1996, the Government of Odisha announced a comprehensive State Agricultural Policy wherein agriculture was given the status of industry. This was done to attract private investment into agriculture. Besides, emphases were laid on expanding irrigation facility to 50% of cultivable land through completion of incomplete projects and promoting individual and group irrigation works, adopting soil management in water-logged, shifting cultivation, waste land and saline and alkaline areas. Programmes were undertaken to identify thrust crops to be promoted in different agro-climatic zones. An agreement was signed with the Agricultural and Processed Food Products Export Development Authority (APEDA) to set up in the export of ginger and turmeric and one Agri Export Zone (AEZ) has been operating in Kandhamal district since 2003. As a continuation to the earlier policy initiative the New Agricultural Policy 2008 has been announced by the state government. The New Policy being futuristic and very flexible it sets the goal in addressing emerging issues, identifying potential areas for development and chalking out a clear agenda for agricultural development for the next 10 years. The broad objectives of the New Policy are

i. To bring in a shift from the present level of subsistence agriculture to a profitable commercial agriculture so that people would accept agriculture as a vocation
ii. To enhance productivity of important crops at least to match with national average through seed replacement, availability of quality planting materials, farm mechanization, technology transfer, etc.
iii. To encourage crop diversification in uplands and medium lands—from paddy to non-paddy crops
iv. To focus on horticultural crops including dry-land horticulture
v. To encourage modern farming system
vi. To increase access to credit for small and marginal farmers
vii. To facilitate appropriate market linkages for agricultural produce with respect to which the state has competitive advantage

Under the present policy, various incentives and subsidies are provided to help farmers in the purchase of hybrid inputs and accessing credit from formal agencies, availing the benefit of network extension in the cultivation of non-traditional varieties and marketing the products. Funds are allocated in improving rural infrastructures particularly roads, godowns and cold storages. Implementation of Seed Village Schemes—a model of participatory production of quality seeds with the involvement of farmers in all the blocks has been implemented. Under the scheme, incentive in the form of subsidy is given in the production of vegetable seeds, oilseeds and pulses seeds. For the development of horticultural crops such as fruits, flowers and vegetables the National Horticultural Mission, a Centrally Sponsored Scheme operating in 24 districts has been extended to six other non-mission districts with state support. To develop marketing facility for agricultural produce the establishment of Krushak Bazar has been introduced, aiming at creating primary rural markets, training farmers and launching awareness campaigns in rural areas. The new policy being adopted recently contains many new programmes and it is too early to evaluate its achievement/failure in the context of crop diversification in the state.

5.4 Conclusion and Policy Implications

The findings of the study amply demonstrate that there have been some changes in the cropping pattern in the state from paddy to non-paddy crops. The growth in area under paddy crop has been negative while that under non-paddy crops has been positive. In the case of some individual crops, the annual growth rate in area has remained high. Inter-regional disparity persists in the extent and pattern of crop diversification. Agro-climatically better-off regions such as the Coastal Plain have not shown better performance in diversifying to HVCs. The small farmers are found positively associated in diversifying to high value crops such as vegetables and fruits. However, inter-regional disparity in this context is noticed. Thus, it can be concluded that crop diversification has taken a slow momentum in the state. However, the small holders have played positive and significant role in diversifying acreages towards fruits and vegetables. No doubt, the state government has taken a number of policy initiatives towards diversifying to HVCs yet they have not yielded desired results. Based on the findings of the study we can suggest that to accelerate the pace of diversification from cereals to HVCs and commercial crops in the state, steps need to be taken to improve rural infrastructure such as irrigation and road connectivity. This will greatly reduce the agro-climatic disparity in growing crops across regions. As the small farmers are found playing a positive role in diversification process, they need to be provided increased institutional support in accessing to better input, credit and marketing their produce.

References

Government of Odisha, Economic survey (Several Issues)

Joshi PK, Birthal PS, Minot N (2006) Sources of agricultural growth in India: role of diversification towards high value crops. MTID Discussion Paper No 98, IFPRI, Washington, DC

Longhurst R (1988) Cash crops in household food security and nutrition. IDS Bull 19(2):28–36

5.4 Conclusion and Policy Implications

The findings of the study clearly demonstrate that there have been some changes in the cropping pattern [illegible]

[illegible]

Chapter 6
Pattern of Agricultural Diversification in Odisha

H. N. Atibudhi

6.1 Introduction

Agriculture continues to remain the predominant sector of Indian economy in terms of employment and livelihood. About 52% of workforce is still engaged in agriculture, even though its share in gross domestic product (GDP) has declined from over 50% in the early 1950s to 14.2% in 2010–2011. The 11th Five-Year Plan also indicates that agricultural development is an important component of inclusive growth approach. However, during the last one and a half decade Indian agriculture has been facing severe challenges, the most serious being the deceleration in its growth rate from about 3.3% during the period between 1980–1981 and 1994–1995 to around 2% between 1995–1996 and 2008–2009. This has serious implications for food security, farmers' income and poverty. There is widespread rural distress leading to a large number of suicides by the farmers in some parts of the country. The economic reforms have shifted inter-sectoral terms of trade in favour of high value agriculture. To overcome the stagnation, the policy decision was to give the next thrust to the eastern states of Odisha, West Bengal, Bihar, Chhattisgarh and Jharkhand. Attention has to be given for crop diversification to meet the annual requirement of 17 million t pulses against the production of 13.1 million t. As far as oil seeds are concerned, our annual average production of 25 million t has to be increased. We also have to divert for more production of fruits, vegetables, milk, fish, meat and other animal products to meet the upsurge in demand and better value realization by the farming community.

Since the 8th Five-Year Plan in Odisha, efforts are being made to diversify agricultural crops and particularly to divert land from paddy to non paddy crops like pulses, oil seeds, sugarcane and vegetables to ensure better returns to farmers. The State is predominantly agrarian with agriculture and animal husbandry sector together contributing 17.84% to the gross state domestic product (GSDP) (2009–

H. N. Atibudhi (✉)
Department of Agricultural Economics, Orissa University of Agriculture & Technology, Bhubaneswar, Orissa, India
e-mail: hnatibudhi@rediffmail.com

M. Ghosh et al. (eds.), *Diversification of Agriculture in Eastern India*,
India Studies in Business and Economics, DOI 10.1007/978-81-322-1997-2_6

2010) at current prices and providing employment to 65 % of the total workforce (as per 2001 Census), directly or indirectly. Thus, Agriculture plays a critical role in the economy and livelihood of majority of its populace. The geographical area of the state is 155.71 lakh ha and divided into two broad regions; the plateau region and the coastal region. The plateau region comprises 77 % of total geographical area of the state. This region is drought prone and divided into seven agro-climatic zones. The remaining 23 % of geographical area, constituting the coastal region, is divided into three agro-climatic zones. The coastal alluvial soil within 10 km from the sea coast is saline. The lowland suffers from water logging and this region is flood prone. On area basis, the state occupies ninth position in India. The total cultivated land of the state is 61.80 lakh ha, out of which 29.14 lakh ha (47 %) is high land, 17.55 lakh ha (28 %) medium land and 15.11 lakh ha (25 %) low land. About 35 % of cultivated land is irrigated and rest 65 % is rainfed, which is exposed to vagaries of monsoon. Rice is the principal crop, which occupies 67 % of the cultivated land in kharif season. However, cropping during rabi season is confined to the irrigated tracts and land with available residual moisture in the soil, which mostly depends on the occurrence of rainfall during the last part of September. The total food grain production in the state has had a fluctuating trend with small growth rates since 2002–2003. In 2008–2009, it was 73.93 lakh t, sliding down from 81.44 lakh t in 2007–2008. Paddy production constitutes more than 90 % of total food grain production in the state. In terms of acreage, there has been gradual shift from paddy to cash (i.e. commercial) crops and this trend continues, but paddy remains to be the dominant crop. As agriculture is the mainstay of Odisha's economy, any fluctuation in the output in this sector profoundly affects the growth rate of the state's economy. The secondary sector is yet to make any significant dent in respect of output and employment generation. Consequently, in a bad agricultural year, there is deterioration in the economic status of the population, particularly in rural areas. The low economic status of the people of the state in rural areas is due to low agricultural productivity. Therefore, diversification of agriculture was very often stressed by the policy makers for sustainable development.

In the agricultural context, diversification can be regarded as the re-allocation of some of a farm's productive resources, such as land, capital, farm equipment and paid labour, to new activities. These can be new crops or livestock products, value-adding activities, provision of services to other farmers and particularly in richer countries, non-farming activities such as restaurant and shops. Factors leading to decisions to diversify are many, but include; reducing risk, responding to changing consumer demands or changing government policy, responding to external shocks and more recently, climate change.

The term 'diversification' has been derived from the word 'diverge' which means to move or extend in the direction different from a common point (Jha et al. 2009). Agricultural diversification can be described in terms of the shift from the regional dominance of one crop towards the production of a large number of crops to meet the increasing demand of those crops. It can also be described as the economic development of non-agricultural activities (Start 2001). The process of diversification can be classified into horizontal and vertical diversification. Horizontal diversification can be referred to as that form of diversification wherein farmers diversify their agricultural activities in order to either stabilize or increase their income or both. It can either take the form of shift from subsistence farming to commercial farming or the shift from low value food crops to high value crops. Vertical diversification refers to the farmers' access to non-farm income, i.e. the income from non-agricultural sources (Haque 1996).

Since 83 % of the farmers in Odisha are small and marginal farmers and their income from crop cultivation as well as non-farm income is not enough to meet their subsistence level. The state produces only a few specialized crops and it is more prone to risk due to uncertain weather conditions, fluctuations in domestic and international prices. Hence, both the horizontal and vertical diversification become the need of the hour. This study aims at analyzing the trends and patterns of agricultural diversification in Odisha. The main objectives of this study are: (i) to analyse the trends and patterns of rural diversification, including both horizontal and vertical diversification in Odisha and (ii) to identify potential sources and suggest appropriate strategies and policies for future sustainable agricultural growth in the state.

6.2 Data Base and Methodology

Secondary data have been used for this study. The secondary data at district, state and national levels were collected from different departments/agencies/publications relating to different variables/parameters of the study for different period. Most of the secondary sources of data were collected from the Directorate of Economics and Statistics, Department of Agriculture and Cooperation, Ministry of Agriculture, Government of India, National Sample Survey Rounds of Central Statistical Organisation (CSO), State Statistical Abstracts, Department of Animal Husbandry & Dairying, National Horticultural Board and Economic Survey and Government of India.

Agricultural diversification in the state was gauged from share of various subsectors in GDP as well as total value of output from agriculture and allied activities, cropping pattern etc. Further index of crop diversification has been estimated by using Simpsons Index, which indicates extent of diversification.

6.3 Results

6.3.1 Trends and Patterns of Agricultural Diversification

Agriculture and allied activities including crop and animal husbandry, fisheries, forestry and agro processing provides the basis of our food and livelihood security. Agriculture and allied activities also provide significant support for economic growth and social transformation of the country. As one of the world's largest agrarian economies, the agriculture sector (including allied activities) in India accounted

Table 6.1 Share of gross state domestic product from agriculture and allied activities. (Source: Central Statistical Organisation, National Accounts Division)

Year	Agriculture		Non-agriculture	
	Odisha	India	Odisha	India
1980–1981	50.19	35.70	49.81	64.30
1985–1986	49.21	31.71	50.79	68.83
1990–1991	36.56	29.28	63.44	70.72
1995–1996	40.87	26.49	59.13	73.51
2000–2001	27.69	23.35	72.31	76.65
2001–2002	29.21	23.20	70.79	76.80
2002–2003	27.43	20.87	72.57	79.13
2003–2004	29.39	20.97	70.61	79.03
2004–2005	25.36	19.20	74.64	80.80
2005–2006	24.63	18.80	75.37	81.20
2006–2007	24.06	18.35	75.94	81.65
2007–2008	22.97	18.11	77.03	81.89

for 14.20 % of the GDP and contributed approximately 10 % of total exports during 2010–2011. The share of this sector in Odisha is 17.84 %. Notwithstanding the fact that the share of this sector in the GDP has been declining over the years, its role remains critical as it provides employment to around 65 % of the workforce.

6.3.2 *Share of Gross State Domestic Product from Agriculture and Allied Activities*

Table 6.1 shows the changes in the shares of agriculture vis-à-vis non-agriculture in the state domestic product in Odisha and in the country. It may be seen from the table that in the country as a whole, the share of agriculture in the GDP declined from 35 % in 1980–1981 to 18.11 % in 2007–2008, while that of non-agriculture increased from 64.30 to 81.89 % in the same period. The same trend was also marked in Odisha. In this state, the share of agriculture in GSDP is higher than the all India level. Conversely, in the state, the share of GSDP from non-agriculture is lower than that of all India level (Table 6.2).

In the state the GSDP declined from 50.19 % in 1980–1981 to 22.97 during 2007–2008, whereas, the contribution of non-agricultural sector increased from 49.81 % in 1980–1981 to 77.03 % in 2007–2008. The contribution of non-agricultural sector was less in the state in comparison to contribution of the sector at the all India level. This indicates the importance of the sector and warrants special attention to protect the income level of the people dependant on this sector.

6.3.3 *Crop Diversification*

There has been a significant change in the cropping pattern as well as in the relative share of various crops in the gross value of crop output in the past few decades.

Table 6.2 Share of GDP within agriculture and allied activities in Odisha and India (%)

Year	Odisha			India		
	Agriculture including livestock	Forestry and logging	Fishing	Agriculture including livestock	Forestry and logging	Fishing
1980–1981	88.07	9.44	2.49	91.89	6.17	1.94
1985–1986	88.38	8.41	3.21	91.36	6.06	2.57
1990–1991	81.33	13.57	5.10	91.46	5.47	3.07
1995–1996	88.89	6.50	4.61	91.81	4.25	3.94
2000–2001	83.03	9.78	7.19	90.96	4.29	4.75
2001–2002	84.42	8.71	6.87	90.93	4.30	4.78
2002–2003	85.11	7.74	7.15	90.14	4.46	5.40
2003–2004	86.16	8.02	5.82	90.74	4.20	5.06
2004–2005	86.66	7.40	5.94	90.77	4.23	5.01
2005–2006	86.84	7.19	5.97	90.46	4.20	5.34
2006–2007	86.41	8.05	5.53	91.24	3.86	4.90
2007–2008	86.73	7.56	5.70	91.78	3.71	4.50

Table 6.3 Simpson's Index of diversification. (Source: Calculated based on the data collected from DE&S Ministry of Agriculture, Government of India)

Regions	1970–1971	1975–1976	1980–1981	1985–1986	1990–1991	1995–1996	2000–2001	2005–2006	2007–2008
India	0.17	0.16	0.18	0.18	0.21	0.24	0.25	0.27	0.26
Odisha	0.31	0.33	0.41	0.45	0.29	0.37	0.40	0.39	0.39

The SID was compared taking two time periods, 1970–1971 and 2007–2008, or the latest year for which data was available. In Odisha, diversification away from food grains was more throughout compared to the country over the years. In Odisha around 39 % diverted for non-food crops whereas in the country it was around 26 % in 2007–2008 (Table 6.3). Horizontal diversification beyond certain point may not be feasible because of the compulsion of the farmers to grow basic food grains for their won consumption needs and non-problematic market sales.

6.3.4 *Changes in Cropping Pattern*

In Odisha, the area share of cereals in the GCA declined from 68.33 % in 1970–1971 to 51.35 % in 2007–2008. Proportionate area under pulses showed substantial increase. It increased from 0.85 % in 1970–1971 to 9.53 % in 2007–2008.The area under oilseeds increased from 4.16 % in 1970–1971 to 8.81 % in 1995–1996 and thereafter declined to 3.58 % in 2007–2008.This could be due to policy of cheap imports of edible oil at all India level and closure of oil fed, the important government promoted organization for oil seed production and processing in the state. There has not been any significant change in areas of fruits and vegetables. However area under sugarcane and jute declined in the state. Thus, it is seen from the analysis of cropping pattern that though area diversification away from food grains has oc-

Table 6.4 Cropping pattern in Odisha and India (% of GCA)

Crops	1970–1971	1975–1976	1980–1981	1985–1986	1990–1991	1995–1996	2000–2001	2005–2006	2007–2008
India									
Cereals	70.3	70.14	69.97	68.79	66.42	64.46	64.04	61.8	61.89
Pulses	12.75	13.38	12.41	13.12	12.81	11.87	10.77	11.38	11.87
Oilseeds	9.41	9.26	9.72	10.22	12.55	13.83	12.05	14.16	13.20
Cotton	4.30	4.02	4.32	4.05	3.87	4.82	4.51	4.41	4.69
Jute and mesta	0.61	0.50	0.72	0.81	0.53	0.50	0.54	0.46	0.48
Coconut	0.59	0.59	0.6	0.66	0.77	0.97	0.97	0.99	1.05
Sugarcane	1.48	1.51	1.47	1.53	1.92	2.21	2.29	2.13	2.51
Fruits and vegetables	0.30	0.41	0.54	0.60	0.64	0.80	1.38	1.56	1.59
Spices	0	0	0	0	0	0	0.89	0.77	0.67
Odisha									
Cereals	68.33	66.19	57.42	53.44	48.71	52.9	52.00	51.66	51.35
Pulses	0.85	1.14	1.71	1.97	21.84	9.62	8.00	9.06	9.53
Oilseeds	4.16	4.19	6.94	8.68	7.73	8.81	3.50	3.71	3.58
Fruits and vegetables	0.37	0.09	0.10	0.44	0.16	0.18	0.28	0.30	0.30
Sugarcane	0.55	0.59	0.60	0.62	0.51	0.51	0.21	0.18	0.22
Jute	0.65	0.49	0.54	0.6	0.37	0.27	0.35	0.28	0.31

curred in the state, the diversification towards horticultural crops has not been much over the years in the state (Tables 6.4 and 6.5).

6.3.5 Share of Value of Crops from Different Crop Production

It is interesting to note that over the year the share of value of output of cereals remained between 30 and 40 % whereas the fruits and vegetable sector contributed from 19 to 52 % over the period. The other important contributor in value term of the output in the state includes spices and condiments which contribute around 4–5 % of the value. To increase the income level of the farmers engaged in agriculture it is not only important to increase in area, production and productivity from fruits and vegetable but also to create facilities for value addition, post harvest management and processing of the produce for better income realization.

6.3.6 Contribution of Allied Sectors

The share of livestock in agriculture and allied activities was 23.86 % in TE-2002 which increased to 25.24 % in 2006 in all India level. Table 6.6 shows that in the state of Odisha, the value of livestock sector contributed only 11.86 % of the total agriculture and allied activities during TE-2002 which increased to 13.62 % in

Table 6.5 Share of value of output from different types of crop production (%)

Crops	1980–1981	1985–1986	1990–1991	1999–2000	2000–2001	2001–2002	2002–2003	2003–2004	2004–2005	2005–2006
Odisha										
Cereals	36.49	36.75	32.77	33.54	30.65	39.40	23.97	38.07	36.89	37.06
Pulses	13.72	11.71	11.68	0.03	0.03	0.03	0.03	0.03	0.03	0.04
Oilseeds	11.44	12.46	14.10	2.51	2.05	2.06	2.18	2.46	2.74	2.76
Sugarcane	3.46	3.41	3.14	1.14	0.95	0.53	0.81	0.72	0.71	0.87
Fibres	1.20	1.06	1.15	0.79	0.81	0.60	0.76	0.87	1.06	1.28
Condi-ments and Spices	4.95	4.29	4.09	5.43	3.93	3.41	4.74	3.68	3.68	3.58
Fruits and vegetables	19.35	19.58	21.90	42.22	47.73	41.48	52.71	41.49	42.42	41.43

Table 6.6 Changes in the share of livestock (TE-2002 to TE-2006) (%). (Source: CSO, Government of India (2008))

States	TE-2002	TE-2006
Odisha	11.86	13.62
India	23.86	25.24

2006. The contribution in value terms of the sector in the state was abysmally low compared to all India level and need due emphasis is in the state policy (Tables 6.7 and 6.8).

The contribution of forestry to the GSDP in agriculture has declined in all India level from 4.20 to 3.75 % in 2006. However, there has been no significant change in the share of the sector in the state to GSDP in agriculture. It continued to contribute 8.19 % to the GDP of agricultural and allied activities.

Fisheries continue to be one of the very important sub-sectors in Odisha agriculture. The sector in value terms contributed around 5.95 % of agricultural GDP in the TE-2002 and 6.27 % in the TE-2006, which was higher than the all India average of 4.32 %. This indicates the significance of the sub-sector in the state for diversification to allied activities.

6.4 Conclusions and Policy Implications

In the country as well as in the state the share of GDP from agriculture has declined, while that of non-agriculture increased significantly in the past few decades. There has been a significant change in the cropping pattern in the country as a whole as well in Odisha in particular during the period. The area share of cereals in the GCA has been highest amongst other crops from 1970–1971 to 2007–2008. It was also observed that the area devoted to food grains (cereals and pulses) was much higher in both state and as well as in the country level.

Table 6.7 Share of forestry and logging in the gross state domestic product (%). (Source: CSO, Ministry of Statistics and Programme Implementation, Government of India)

States	TE-1983	TE-2002	TE-2009
Odisha	5.74	8.56	8.19
India	6.38	4.20	3.75

Table 6.8 Share of fisheries sub-sector

States	TE-2002	TE-2006
Odisha	5.95	6.27
India	4.15	4.32

From the Simpson's Index of crop diversification (SID), it was seen that diversification away from food grains in Odisha was more in comparison to all India level but reduced especially since 1995–1996. In Odisha, maximum diversification of crop was towards pulses and oilseeds but after 2000, the area under oil seeds has been declining. There have also been significant changes in the relative shares of various crops in the gross value of crop output from agriculture (crop sector) in the past few decades. Specifically the contribution of fruits and vegetables are significant in the crop sector. There is a need to sustain the contribution of forestry and fisheries seeing the importance of these sectors for the state economy. The uncertainty of the crop production sector has warranted special policy interventions to strengthen the ailing livestock sector, which can provide supplementary incomes to the farmers and can contribute to the GSDP.

Despite the fact that farmers are diversifying horizontally more within the crop sector compared to the allied sectors, they need to diversify much more towards high value crops and also within the allied sectors.

The state is blessed with varied agro-climatic conditions suitable for growing a variety of fruits, vegetables and spice crops. Hill tracts of KBK (Kalahandi, Bolangir and Koraput), Kandhamal and Gajapati districts are conducive for horticultural activities. The state must exploit its potentialities to diversify for this high value crops through National Horticultural Mission to realize the benefit.

Further vertical diversification towards non-farm activities was also very less in the state. An integrated policy support system is required for promoting sustainable horizontal and vertical diversification of the rural economy in the state.

The fisheries and agro-forestry sector contributes significantly in the state agrarian economy. As the state is facing natural calamities every alternate year, the allied sectors need special attention by the Government. The fisheries sector was mostly constrained by lack of timely credit, inaccessibility to cold storages, poor road conditions and connectivity and transportation problems. Thus, in the crop, livestock and fisheries sector policies have to be made such that the specific problems faced by farmers can be mitigated.

Further, in the state 83 % of farmers are small and marginal farmers and basically they are part-time farmers. The idle labour from these farm categories in off season can be diverted for vertical diversification for agro based industries, agri-business, agro-processing and other services sector. The state can explore agro-based tourism to diversify and boost economic conditions of rural people.

References

Haque T (1996) Small farm diversification: problems and prospects. NCAP, New Delhi

Jha BN, Kumar, Mohanty B (2009) Pattern of agricultural diversification in India, Institute of Economic Growth, working paper no E/302/2009

Start D (2001) Rural diversification: what hope for the poor? Summary themes from Start and Johnson 2001: transformation, well being and the state. Rural livelihood diversification in South Asia, ODI working paper

Chapter 7
A Study on the Extent of Crop Diversification in West Bengal

S. Maji, A. Jha Chakraborty, B. K. Bera and A. K. Nandi

7.1 Introduction

Crop diversification is considered as an important risk management tool for growth augmentation, farm income stabilization, employment generation, conservation of natural resources, poverty alleviation and export promotion. The shift in area from one crop to another or crop substitution is also necessary to meet the demand for diversified products including high value food commodities arising out of the changing consumption patterns of consumers. The experiences of many developing countries have revealed the changes in production portfolio at the farm level due to altering dietary patterns (Barghouti et al. 2003; Dorjee et al. 2002). With the greying of the green revolution, marginal growth in agriculture particularly during the post-liberalization period has become an area of great concern to policy makers. At the same time, low productivity growth of food grains coupled with increasing input prices has rendered farming non-remunerative. This has prompted the farming community to diversify production portfolio towards high value food commodities (Joshi 2005). There was substantial crop area shift from coarse cereal and pulses to rice, wheat, sugarcane and oilseeds during the 1970s and 1980s and in the subsequent period from crop sector to high value fruit-vegetable crops, forestry, livestock and fishery activities (Sawant and Achuthan 1995; Vyas 1996; Joshi et al. 2004). Diversification of crops is the consequence of many factors most of which are interrelated with each other, such as slow growth, low farm income,

B. K. Bera (✉) · S. Maji · A. Jha Chakraborty · A. K. Nandi
Department of Agricultural Economics, Bidhan Chandra Krishi Viswavidyalaya, Mohanpur, Nadia, West Bengal, India
e-mail: bkbera_econ@ rediffmail.com

S. Maji
e-mail: sudeshnamaji08@gmail.com

A. Jha Chakraborty
e-mail: agninomics@gmail.com

A. K. Nandi
e-mail: aknandibckv@rediffmail.com

M. Ghosh et al. (eds.), *Diversification of Agriculture in Eastern India*,
India Studies in Business and Economics, DOI 10.1007/978-81-322-1997-2_7

change in consumption pattern, etc. Besides, the expanding urbanization, increasing infrastructural development and liberalization of trade policies have been identified as factors that triggered the process of agricultural diversification out from an area of stable food production (Joshi et al. 2004; Pingali 2004). Apart from economic factors, agro-climatic situations and their change over time also influence shift in area and crop substitution. Crop diversification and the growing of large number of crops are practiced in rainfed lands to reduce the risk factors due to drought or less rain and in areas with distinct soil problems (Hazra 2001). Whatever, may be the reasons, the fact is that Indian agriculture has been diversifying during the last two decades towards high-value commodities (HVCs) i.e. fruits, vegetables, milk, meat and fish products and accelerated during the decade of 1990s (Bhattacharya 2008).

West Bengal, being endowed with a diversified agro-climatic situation, is suitable for growing wide range of crops. However, the state has been witnessing crop diversification in recent years, and that has remained confined mostly to food grains, that is, shifting of area from autumn rice and winter pulses to summer rice. But very recently, increase in food grains productivity has made it possible to allocate more area towards other crops such as oilseeds, with a severe supply shortage, and thus the specialization tendency witnessed earlier has given room for overall crop diversification (Bhattacharyya 2008). However, the shift in area from food crops to non-food crops, is slow with rates declining in the whole post-liberalization period. The Rate of change in the cropping pattern in favour of high-value horticultural crops or oilseeds is not as expected, particularly, after the opening up of Indian economy in 90s. In a move to promote diversification, the state has prepared an action plan that includes limiting of rice acreage without hampering total production through the introduction of high-productivity hybrid rice and the providing of an incentive package to growers for raising acreage under pulses, oilseeds, maize and wheat, in which the state is deficient in production.

Against this backdrop, the present study proposes to examine the nature and extent of diversification in agriculture sector of West Bengal over the period 1965–1966–2007–2008. The specific objectives of the study are to (1) study the inter-temporal change in cropping pattern of West Bengal; (2) find out the extent of diversification in the state; and (3) examine the magnitude of expansion and substitution effect on diversification.

7.2 Materials and Methods

The study is based on secondary data collected from various issues of Statistical Abstract published by the Bureau of Applied Economics and Statistics, Government of West Bengal as well as the Economic Review and Estimates of Area, Production and Productivity of Principal Crops in West Bengal published by Government of West Bengal. The reference period for the study is 1965–1966 to 2007–2008. The extent of diversification has been measured using methods of Herfindahl Index,

Ogive Index, Entropy Index, Modified Entropy Index and Composite Entropy Index at three points of time, that is, 1967–1968, 1987–1988 and 2007–2008.

Herfindahl Index is computed as follows

$$\text{H.I.} = \sum_{i=1}^{N} = P_i^2$$

where N = total number of crops; P_i = acreage proportion of the i-th crop in total cropped area. The value ranges between zero to one. Since the index is a measure of concentration, it was transformed by subtracting it from one that is, $1 - \text{H.I.}$ The transformed value of H.I. will avoid confusion to compare it with other indices. It takes zero value in case of complete diversification and value one indicates perfect specialization.

Ogive Index is measured as follows:

$$\text{O.I.} = \frac{\sum_{i=1}^{N} \left\{ P_i - \left(\frac{1}{N} \right) \right\}}{\left(\frac{1}{N} \right)}$$

Ogive Index is also a measure of concentration. It is transformed as $1 - \text{O.I.}$ It implies that the index approaches zero in extreme cases of perfect concentration as well as perfect diversification.

Entropy Index is computed as:

$$\text{E.I.} = -\sum_{i=1}^{N} P_i \log P_i$$

or

$$\text{E.I.} = -\sum_{i=1}^{N} P_i log \left(\frac{1}{P_i} \right)$$

This index will increase with the increase in diversification and it approaches zero when there is perfect concentration. The Entropy Index has the limitation of not giving standard scale in assessing the degree of diversification and to overcome the limitation Modified Entropy Index is used. Modified Entropy Index is estimated by using following equation:

$$\text{M.E.I.} = -\sum_{i=1}^{N} P_i log_N P_i$$

or

$$\text{M.E.I} = \frac{\text{E.I.}}{\log N}$$

The value of Modified Entropy Index ranges between zero and one. This index is not sensitive to the change in the number of crops. If there is specialization, then the index takes the value zero and one in case of diversification.

The formula of calculating Composite Entropy Index is given by:

$$\text{C.E.I} = -\left(\sum_{i=1}^{N} P_i \log_N P_i\right)\cdot\left(1-\frac{1}{N}\right)$$

or

$$\text{C.E.I.} = \text{M.E.I.}\cdot\left(1-\frac{1}{N}\right)$$

The value of the Composite Entropy Index increases with the decrease in concentration and rises with the number of crops or activities. Both components of the index are bounded by zero and one, and thus the value of the Composite Entropy Index ranges between zero and one.

Log linear growth rate technique is applied to measure annual growth rate of area of each individual crop taking in to consideration the period ranging from 1977–1978 to 2007–2008. The equation of log linear growth rate is given below:

$\ln Y = \ln a + t \ln b + u$ where Y = area under crop, t = time period and u = error term, and a and b are the parameters to be estimated.

Once the rates of growth and the change in the rates of growth of the acreage of different crops are known, the substitution effect and expansion effect can be estimated easily. For a given gross cropped area (GCA), the substitution effect is defined as the relative decline in area under some crops and the compensating increase in area of the substitutable crops. On the other hand, the expansion effect is defined as the expansion of GCA. First of all to ensure whether the area under any crop has been changed due to the inter-crop shift of area or because of change in the total area under cultivation, a simple elasticity measure is done.

The elasticity measure (E) defined by Venkataraman and Prahladachar (1980) is used in this paper. The value of E under each crop is calculated to identify the crops that have gain in area from other crops and those which lose area to the former. If E is found to be greater than unity for any crop, it can be said that the area under that crop has increased due to both substitution effect and expansion effect. On the other hand, if $E<0$ for any crop, it can be easily asserted that the crop has lost area to crops having elasticity greater than unity. However, if the value of E is such that $0<E<1$, it is difficult to say whether the rise in area is due to expansion of area or due to substitution effect. In this case, it can only be said that the area of the crop has increased at a rate less than that of GCA.

Table 7.1 Index number of area under principal crops or crop groups of West Bengal for the period from 1965–1966 to 2007–2008 (Base: Triennium ending crop year 1967–1968 = 100)

Crops	Period	
	TE 1987–1988	TE 2007–2008
Aus	87.90	43.20
Aman	102.38	100.97
Boro	1763.47	3817.52
Total rice	113.68	122.66
Wheat	613.36	609.31
Total cereal	119.05	127.56
Total pulses	50.28	28.43
Total food grains	109.79	114.21
Rapeseed & mustard	285.30	393.61
Total oilseed	376.76	437.27
Jute	126.30	133.30
Total fibres	129.30	125.09
Potato	207.87	872.75
Total vegetables	381.59	499.51
Total fruits	1074.25	1735.81

7.3 Result and Discussion

At the outset, we will examine the percentage change in area under various crops individually or in groups at three points of time to measure inter-temporal variations in crop acreage. The temporal change in area share by major crops or crop groups is presented in Table 7.1.

Table 7.1 reveals that the share of food grains area has witnessed an acceleration in both periods by 9.79 and 14.21 %, respectively in respect to base period, that is, the triennium ending 1967–1968 in spite of experiencing an windfall setback in total pulses area due to the persistent rise in area under total cereals. Both total rice and wheat, comprising total cereals, are found to have risen successively, but wheat area has grown at a faster rate compared to rice. Among rice crops, area under aman registered a marginal rise by only 2.3 and 0.97 % in the triennium ending 1987–1988 and 2007–2008, respectively. Windfall gain in boro area has not only compensated the drastic fall in area under aus rice but also pushed the share of total rice to rise constantly. It appears that area left by total pulses and aus has been largely, if not fully, occupied by boro rice and wheat. Again, the shift in area from one crop to another has mostly taken place during the period triennium ending 1967–1968 to 1987–1988 in comparison to later except boro and aus both of which have experienced greater rise and loss in area share respectively during period triennium ending 1987–1988 to 2007–2008. All the non-food grains have maintained a steady rise in area with varying magnitude, in all period, but the percentage increase in area is relatively slow during second half compared to first half except potato which registers a phenomenal rise of 772.75 % during 1987–1988 to 2007–2008 as against to 107.81 % during 1967–1968 to 1987–1988.

Table 7.2 Change in cropping pattern of West Bengal (Triennium ending 1967–1968, 1987–1988 and 2007–2008)

Index	Period		
	TE 1967–1968	TE 1987–1988	TE 2007–2008
Aus	10.29	7.07	2.91
Aman	62.14	49.70	41.15
Boro	0.58	8.07	14.68
Total rice	73.02	64.80	58.75
Wheat	0.91	4.38	3.65
Total cereal	75.80	70.50	63.42
Total pulses	11.78	4.63	2.19
Total food grains	87.59	75.10	65.62
Rapeseed and mustard	1.65	3.69	4.27
Total oilseed	2.44	7.20	7.25
Jute	6.89	6.80	6.02
Total fibres	7.57	7.61	6.42
Potato	1.23	2.00	3.97
Total vegetables	2.82	8.41	9.35
Total fruits	0.16	1.40	1.99

However, change in cropping pattern measured in terms of area under a crop as a percentage of GCA presents a different scenario. At the initial period, 87.59 % of GCA is captured by food grain, of which the share of total rice is accounted to be 73.02 %. With the decline in area under food grain in successive area, the dominance of total rice has come down along with a moderate rise in wheat area whereas the area under total pulses is reduced remarkably. Gradual fall in *aman* and *aus* and a notable increase in *boro* area are quite prominent. During this period, total oilseeds have witnessed a substantial increase in area along with potato, vegetables and fruits. Here also shift in area is a percentage of gross cropped area is higher during 1967–1968 to 1987–1988 in comparison to 1987–1988 to 2007–2008. Although, all non-food grains are found to have been gaining importance in cropping pattern of West Bengal, total cereals still dominates the agriculture. So, it is evident that the state is witnessing a crop diversification in crop sector and food grain area is shifting towards the cultivation of non-food grains particularly, oilseeds, potato, vegetables and fruits, although the rate of change is slow in the later compared to the previous period.

Considering significant cropping pattern changes that have taken place in the state, it is pertinent to examine the extent of crop diversification by applying various diversification index measures. The extent of diversification or specialization is measured by employing diversification index such as Herfindahl Index, Ogive Index, Entropy Index, Modified Entropy Index and Composite Entropy Index. The results of their measures are presented in Table 7.3.

The estimated values reveals an acceleration of crop diversification in the state as the values of all indices taking all crops have risen in successive periods. Herfindahl Index and Ogive Index have gone up from 0.60 and 0.67 in 1967–1968 to 0.73 and 0.81 in 1987–1988 and subsequently moved to 0.78 and 0.85 in 2007–2008

Table 7.3 Diversification indices for West Bengal during 1965–1966 to 2007–2008

Index	Period		
	TE 1967–1968	TE 1987–1988	TE 2007–2008
Herfindahl index	0.60	0.73	0.78
Ogive index	0.67	0.81	0.85
Entropy index	0.59	0.79	0.83
Modified entropy index	0.57	0.76	0.79
Composite entropy index	0.52	0.69	0.73

indicating a gradual shift in cropping pattern. The results of Entropy, Modified Entropy and Composite Entropy Indices also suggest that the state is heading towards crop diversification but at slower pace and the rate of diversification was faster in the first two decades in comparison to later periods. Production of fruits and vegetables is picking up momentum and horticulture is becoming a new option which is being taken up by farmers of quite a few districts (Bhattacharyya 2008).

In this context, it is pertinent to find the sources of crop diversification, that is, whether the change in pattern of crop diversification is due to inter-crop exchange in area (i.e., substitution effect) or change in GCA (expansion effect). An elasticity measurement based on growth rates of area under major crops of West Bengal is applied to delineate the effects on crop diversification. Crops with elasticity value greater than one forms category-I and elasticity values ranging between zero to one and below zero comprise category-II and III respectively. Table 7.4 represents the crops such as summer rice (boro), wheat, total oilseeds including rapeseed and mustard, potato, total vegetables and fruits fall into category-I indicating as gain in area from other crops. Winter rice (aman), total rice, total cereals and total food grains, total fibre including jute comprising category-II reveals indeterminacy as regards the increase in area i. e. whether due to expansion or substitution effect. Category-III consists of crops such as autumn rice (aus) and total pulses ascertain loss in area to other crops. Depending on cropped area, GCA elasticity, the shift in cropping pattern in West Bengal for period TE 1967–1968 to 2007–2008, expressed in terms of expansion and substitution effect is depicted in Table 7.4 (Table 7.5).

It portrays that autumn rice and total pulses have suffered an area setback of 942.80 thousand ha and crops such as winter rice (aman), summer rice (boro), wheat, total cereals, total food grains, total oilseeds, total fibres, potato, vegetables and fruits have gained an area measuring 2613.68 thousand ha. Therefore, the increase in crop area due to substitution effect is 942.80 thousand ha and the remaining 1670.88 thousand ha rise in area can be attributed to expansion effect.

7.4 Conclusions

The agricultural sector of West Bengal is gradually undergoing diversification in favour of high-valued food crops, mainly potato, oilseeds, fruits and vegetables, but the pace of diversification is not as fast as necessary for speeding up of growth. The

Table 7.4 Cropping pattern changes in West Bengal: substitution and expansion effect TE 1967–1968 to 2007–2008

Crops	Substitution effect (in thousands of hectares)	Crops	Substitution and expansion effect (in thousands of hectares)
Aus	−453.17	Aman	91.80
Total pulses	−489.63	Boro	1376.77
		Total rice	1015.50
		Wheat	277.63
		Total cereal	1263.40
		Total food grains	773.20
		Rapeseed and mustard	304.93
		Total oilseed	543.30
		Jute	91.87
		Total fibre	72.50
		Potato	309.23
		Total vegetables	731.84
		Total fruits	183.61

Table 7.5 Crop area—gross cropped area (GCA) elasticity in West Bengal during TE 1967–1968 to 2007–2008

Crops	Elasticity
Aus	<0
Aman	0–1
Boro	>1
Total rice	0–1
Wheat	>1
Total cereal	0–1
Total pulses	<0
Total food grains	0–1
Rapeseed and mustard	>1
Total oilseed	>1
Jute	0–1
Total fibre	0–1
Potato	>1
Total vegetables	>1
Total fruits	>1

cropping pattern is turning against total pulses and autumn rice. It is also evident that in spite of shifting in area from total food grains to high-value horticultural crops and oilseeds, cereals still dominate agriculture sector of West Bengal. The change in cropping pattern is 36.07 % of GCA due to the substitution effect whereas 63.93 % is due to expansion effect. As crop diversification is the prerequisite for economic development in an agriculture dominant economy, increase in irrigation, development of post harvest marketing, processing and infrastructural facilities coupled with remunerative prices is essential to expedite the process of cropping pattern change in the state.

References

Barghouti S, Kane S, Sorby K (2003) Poverty and agricultural diversification in developing countries. The World Bank (Mimeo), Washington, DC

Bhattacharyya R (2008) Crop diversification: a search for an alternative income of the farmers in the state of West Bengal in India. In: Proceedings of the international conference on applied economics (ICOAE), Kastoria, Greece, 15–17 May, pp 83–97

Dorjee K, Broca S, Pinpali P (2002) Diversification in South Asian agricultural trends and constraints. International workshop in agricultural diversification in South Asia, MOB-NCAP-IFPRI, Paro

Hazra CR (2001) Crop diversification in India. Crop diversification in the Asia Pacific region, FAO Corporation Document Depository, RAP Publication 2000/14

Joshi PK (2005) Crop diversification in India: nature, pattern and drivers. International Food Policy Research Institute, IFPRA—South Asia, New Delhi

Joshi PK, Gulati A, Barthal PS, Tiwari L (2004) Agricultural diversification in South Asia: pattern, determinants and policy implications. Econ Polit Wkly 39(24):2457–2468

Pingali PK (2004) Agricultural diversification: opportunity and constraint, FAO Rice Conference, Rome Italy, 12–13 Feb

Sawant SD, Achuthan VC (1995) Agricultural growth across crops and regions: emerging trends and patterns. Econ Polit Wkly 30(12):A2–A13

Venkataramanan LS, Prahladachar M (1980) Growth rates and cropping pattern changes in agriculture in six states: 1950 to 1975. Indian J Agric Econ 35(2):71–84

Vyas VS (1996) Diversification in agriculture—concept, rationale and approach. Indian J Agric Econ 51(4):636–643

References

[illegible]

Chapter 8
A Study of Diversification of Katarni to HYV Paddy in Bihar

Basant Kumar Jha and Rajiv Kumar Sinha

8.1 Introduction

With about 45 million hectares (MHs) of rice area, India is the second largest producer of rice in the world, after China. The rice production in India reached 91.61 million tonnes (MTs) during 2001–2002. The production of rice increased to 95.68 MTs in 2007–2008. Of these, the three states, viz.; West Bengal, Uttar Pradesh and Andhra Pradesh accounted for about 42 % of the total production in the country (Mallick and Roy 2003). India also produces some of the best quality rice grains of the world. These include the 'long-grained export quality Basmati' and a host of 'locally adopted small- and medium-grained scented rice varieties' known for their excellent cooking and eating qualities (Rani and Singh 2003). Most rice growing states in the country have a few or many such varieties. While Basmati is found in Punjab, Haryana and western Uttar Pradesh, the small- and medium-grained scented rice varieties like kalanamak, shakarchini and hansraj are found in Uttar Pradesh; ubraj, chinoor in Chattisgarh; kalajoha in north-east; ramdhuni pagal in Orissa; ambemohar in Maharashtra and so on (Singh et al. 2003). While the long-grained Basmati rice grains are generally exported and have assured markets, the small- and medium-grained non-basmati scented rice grains are consumed locally (Singh and Singh 2003).

In Bihar, although aromatic rice is grown all over the state, it is mainly concentrated in Bhagalpur and Magadh divisions. Bhagalpur has been a traditional aromatic rice growing area, where the varieties, such as katarni, tulsi manjari, badshahbhog, Br-9 and Br-10 are mostly common. These are photoperiod-sensitive, tall and hence, susceptible to lodging and several diseases and pests. Their yields vary from 2.0–2.5 t/ha. Katarni is the most prevalent variety of the region. However,

R. K. Sinha (✉) · B. K. Jha
Agro-Economic Research Centre for Bihar and Jharkhand, Old P.G. Campus,
T.M. Bhagalpur University, Bhagalpur 812007, Bihar, India
e-mail: rajiv.sinha1959@gmail.com

B. K. Jha
e-mail: basantkrjha@hotmail.com

M. Ghosh et al. (eds.), *Diversification of Agriculture in Eastern India*,
India Studies in Business and Economics, DOI 10.1007/978-81-322-1997-2_8

over the period, there has occurred a large variation, which has resulted into various types, such as bhauri katarni, deshla katarni and sabour katarni (Singh et al. 2003). In Magadh region, which is the main rice growing tract of Bihar, farmers grow karibank, marueya, mehijawain, shyamjira, tulsiphool, sonachur and shah pasand. Over the time, the areas under these varieties have drastically reduced, although farmers still grow karibank and marueya, but on a small scale only. The tarai region of West Champaran was, at one time, known for its good quality aromatic rice varieties, that included lal champaran basmati, bhuri champaran basmati, kali champaran basmati, baharni, badshahbhog, chenaur, dewtabhog, kesar, kamod, kankjeera, marcha, ram jawain, sonalari and tulsi pasand. Most of these varieties have been either already lost, or are at the verge of extinction (Thakur et al. 2003).

In view of the above frigid situation in regard to farmers' willingness/response to undertake cultivation of katarni paddy, particularly in the light of declining areas under it, and its diversification towards other High Yielding Variety (HYV) paddy, an attempt has been made in this chapter with two specific objectives. These are: (1) to analyze the level and causes of diversification of area under katarni paddy towards HYV paddy, and (2) to suggest diagnostic measures for sustaining and increasing the area under aromatic katarni paddy.

8.2 Sampling Method and Data Analysis

The chapter is based on the primary data collected from 25 katarni paddy growers of Jagdishpur block under Bhagalpur district of Bihar. It constituted seven marginal farmers, ten small farmers, five medium farmers and three large farmers. The respondents surveyed were randomly selected from potential villages of Jagdishpur block, viz., Jagdishpur, Tagepur and Bhawanipur.

Further, to address the objectives of the chapter, data containing cropping pattern of the surveyed farmers were collected for two points of time, i.e., one–two decades before (1991–1992) and for present time, i.e., for 2010–2011. The diversification/shift of area under katarni paddy towards other crops has been estimated by using simple percentage method.

8.3 Socioeconomic Profile of Respondents

Out of the total 25 growers, 7 belonged to marginal, 10 small, 5 medium and 3 large farmers' groups. Out of the total farmers surveyed, the highest number of respondents (56 %) belonged to other backward class (OBC) group followed by schedule caste (32 %) and general category (12 %) presented in Table 8.1. In scheduled castes (SC) group of farmers, marginal were highest in number (four). No scheduled tribes (ST) farmer was found in the surveyed villages. Among OBC respondents, highest number of farmers (six) belonged to small-farm-size group equally followed by

Table 8.1 Socioeconomic profile of the respondents

Social group	Marginal	Small	Medium	Large	Overall
SC	04	03	01	–	08 (32.00)
ST	–	–	–	–	–
OBC	03	06	03	02	14 (56.00)
General	–	01	01	01	03 (12.00)
Total	07	10	05	03	25 (100.00)

Figures in parentheses indicate the percentage over total

marginal and medium (three) and large (two) only. Out of the overall total of three farmers belonging to social group of general category, each one belonged to small, medium and large-farm-size groups.

8.4 Diversification: An Empirical Peep

With the view to understand the diversification of area under katarni paddy towards HYV paddy or other crops, Tables 8.2 and 8.3 could be looked into intently. The tables contain data showing cropping pattern for the years 1991–1992 and 2010–2011. A perusal of the tables suggests that the area under katarni paddy, owned by the surveyed farmers taken together, which was 9.92 ha in 1991–1992 had declined to 6.70 ha in 2010–2011. It means, there is a decline of 32.46 % in the area of katarni paddy over the 20-years' period. The tables further reveal that the area under paddy (general, including HYV) that was 22.34 ha in the year 1991–1992 for all the surveyed farmers taken together, had increased to 25.56 ha in 2010–2011. It means that there is an increase of 14.42 % in areas under HYV and other paddy over the same period. This implies that some areas (about 18.04 %) under Katarni paddy have also shifted towards other crops like maize, pulses, potato, etc.

8.5 Problems in Growing Katarni Paddy

In regard to the prospects of paddy (in general), and katarni paddy (in particular), it can be said that if the farmers of Bhagalpur and Banka districts are facilitated with ensured, effective and sufficiently desired irrigation system, then this region can produce such a large quantum of paddy and other cereals, that will be hardly stored in godowns (Kumar 2012). Whichever crops will be sown here, particularly in the soil of Jagdishpur block under Bhagalpur district, it will give bumper crop. Jagdishpur is such a block in Bhagalpur district, which has a distinct identity in regard to its peasantry. One of the evident reasons relating to its distinct identity is the aromatic katarni that has special characteristics of capacity to provide jovial pleasure of taste and flavor. But unfortunately, it has been struggling hard to preserve and retain its identity in the region (Kumar 2011). Despite the need to preserve katarni,

Table 8.2 Cropping pattern of the sample households during 1991–1992

Crops	Irrigated		Un-irrigated		Total	
	Area (in ha)	Production (in q)	Area (in ha)	Production (in q)	Area (in ha)	Production (in q)
Paddy (gen+HYV)	22.34	402.12	36.85	737.00	59.19	1139.12
Katarni paddy	9.92	158.72	4.48	67.20	14.40	225.92
Wheat	20.00	360.00	–	–	20.00	360.00
Maize (kharif)	3.63	87.12	6.00	162.00	9.63	249.12
Pulses						
Gram	6.20	49.60	4.66	34.95	10.86	84.55
Masoor (lentil)	–	–	11.30	79.10	11.30	79.10
Moong	–	–	4.00	24.00	4.00	24.00
Khesadi (lethyrus)	–	–	2.02	16.16	2.02	16.16
Oilseeds						
Mustard	1.00	08.00	6.00	48.00	7.00	56.00
Linseed	–	–	3.00	30.00	3.00	30.00
Vegetables						
Potato	1.25	170.00	–	–	1.25	170.00
Onion	0.50	8.50	–	–	0.50	8.50
Brinjal	0.80	10.40	–	–	0.80	10.40
Other						
Orchards (mango)	–	–	5.85	50.00	5.85	50.00

the farmers have not taken any initiative in this regard. The Government of Bihar is also yet to take effective measures in this regard. Agricultural experts opine that if Jagdishpur block under Bhagalpur district is agriculturally developed, then it can turn into heaven of farmers (Mishra 2012).

During the course of survey, in villages of Jagdishpur block under Bhagalpur district, it was reported by most of the respondents that cultivation of katarni paddy in the area is adversely affected by the deepening of the bed of Chandan river. It has been caused due to excess excavation of sand from different points of the river in an uncontrolled way for the last about 8 to 10 years. The farmers of villages situated on both banks of Chandan river have been finding it difficult to grow katarni paddy (in particular) in a substantial area of their total land holdings as irrigation by the river water has turned to be a costly exercise. Further, there is no canal irrigation in the area, and thus, only and main source of irrigation is Chandan river. Despite strong protests by farmers of the area, general public and some public organizations, several tractors and truck loads of sand were being excavated every day.

The problem of irrigation in the area has caused an increasing tendency of repudiation among farmers to undertake any more area, or even maintain the cultivation of katarni paddy in minimum areas that they cultivated during yester years. This is

Table 8.3 Cropping pattern of the sample households during 2010–2011(area in hectare, production in quintal)

Crops	Irrigated		Un-irrigated		Total	
	Area	Production	Area	Production	Area	Production
Paddy (gen+HYV)	25.56	894.60	38.30	1149.00	63.86	2043.60
Katarni paddy	6.70	140.70	3.03	54.54	9.73	195.24
Wheat	18.00	387.00	–	–	18.00	387.00
Maize (kharif)	5.63	152.00	5.00	135.00	10.63	287.00
Pulses						
Gram	–	–	6.00	78.00	6.00	78.00
Masoor (lentil)	–	–	10.00	100.00	10.00	100.00
Moong	–	–	5.30	31.80	5.30	31.80
Khesadi (lethyrus)	–	–	2.00	16.00	2.00	16.00
Oilseeds						
Mustard	1.00	11.00	5.00	50.50	6.00	61.50
Linseed	–	–	2.50	20.00	2.50	20.00
Vegetables						
Potato	2.00	352.00	–	–	2.00	352.00
Onion	0.55	102.30	–	–	0.55	102.30
Brinjal	1.00	221.00	–	–	1.00	221.00
Other						
Orchards (Mango)	–	–	2.00	19.00	2.00	19.00

one of the most drenching and inhibiting constrains that has been leading to decline in the area under katarni paddy. Until and unless this problem is addressed suitably and effectively by construction of check dams, the prospect of katarni paddy production is not much bright, rather dark.

8.6 Prospects

As far as efforts to preserve, sustain and maintain significance (both at national and international levels) are concerned, research under a three-year project is undertaken by the scientists of the Bihar Agricultural University, Sabour, Bhagalpur. Under this research, works are being done related to seed and quality of soil (Thakur 2010). Earlier farmers used to grow katarni paddy by using bio-fertilizers, but since excessive use of chemical fertilizers is being done, its natural taste/fragrance is being eroded (Kumar 2012; Tiwari 2011).

According to respondents, decline in the area under katarni paddy started taking place since 1991–1992. Before that, it was grown for selling purposes, but after 1991–1992, its cultivation was being undertaken with the main objective to maintain social status and giving it as a gift to near and dear ones and VIPs of the society, polity and/or administration.

As a matter of fact, in katarni paddy, the cost of cultivation is comparatively lower. It does not require much chemical fertilizer. Presently katarni paddy is grown in about 200 % of the total land area under cultivation of all varieties of paddy.

In the early 1990s, katarni paddy was grown in nearly 37.00 % areas of the total paddy area in Bhagalpur district of Bihar. But, as consequential effects of frequent floods, after that, particularly the most devastating flood that had occurred in Jagdishpur block of Bhagalpur district and adjoining blocks of Banka district in the year 2001, the area under katarni paddy declined significantly. Further, the deepening of the bed of Chandan river, (i.e., main source of irrigation for the farmers), irrigation has become not only much expensive, but for most of the year, it remains out of the easy access of the farmers too. Therefore, the farmers are compelled to diversify their 'land under katarni paddy' to other varieties of paddy.

8.7 Diagonostic Measures

Katarni paddy, being grown in Bhagalpur and Banka districts of Bihar, is unique and the best in aroma, taste and excellent in flavor. We cannot idle away over the anxiety creating diversification of areas under katarni paddy to other crops without making suitable and possible efforts.

Based on the inputs given by the farmers during the survey, and on the basis of our observation, a few action points have been suggested to save and increase, or at least sustain the area under katarni paddy, which is a *specific largesse of nature to the soil of the* two districts only in the country.

Strict administrative and legal steps should be taken to check uncontrolled excavation of sand from different points of Chandan river flowing through Bhagalpur and Banka districts both. It is highly desirable with the view to assure irrigation water for katarni paddy.

Close vigil need to be kept by agricultural scientists and officers of the Department of Agriculture on the tendency of distorting original katarni paddy seed by some of the big/medium farmers and other traders. Use of such adulterated seed has been posing threat and extenuating original aroma of katarni rice.

The Department of Agriculture, Government of Bihar, Ministry of Agriculture, Government of India, NABARD and other agencies/NGOs working in the field of agriculture and rural development may launch a special campaign to popularize the crisp and most distinguished aroma of katarni rice. It is inevitable with the view to enhance its export potentiality and its demand even in domestic market (various states of the country).

Research needs to be conducted to ascertain, as to which particular micronutrient/(s) is/are responsible for maintaining aroma in katarni paddy. After identification of the same, the specific type of green manure/cattle dung and fertilizer can be appropriately prescribed to sustain that unique fragrance.

Farmers should be warned of using higher doses of chemical fertilizers with a view to get higher yields of katarni paddy. The invaluable aroma of katarni paddy is subjected to be hived off by use of chemical fertilizers.

In view of the lower yield of katarni paddy (nearly 2.5 times lower than the yield of HYV paddy), and with the objective to encourage farmers to have discreet and strong willingness for devoting significant areas to katarni paddy as well, its procurement price by private traders needs to be enhanced.

References

Kumar P (2012, Aug 7) Hindustan, Hindi Daily, Bhagalpur

Kumar R (2011, Nov 29) Dainik Jagaran, Hindi Daily, p 8

Mallik S, Roy TK (2003) The scope of cultivation of aromatic rice in West Bengal. In: Singh RK, Singh US (eds) A treaties on the scented rices of India. Kalyani, Ludhiana, pp 433–446

Mishra, Lalit K (2012, Feb 27) Prabhat Khabar, Hindi Daily, p 1, 17

Rani N, Singh S (2003) Efforts on aromatic rice improvement in India. In: Singh RK, Singh US (eds) A treaties on the scented rices of India. Kalyani, Ludhiana, pp 23–25, 31

Singh RK, Singh US (2003) Scented rice of India: an introduction. In: Singh RK, Singh US (eds) A treaties on the scented rices of India. Kalyani, Ludhiana, pp 1–4

Singh RK, Mani SC, Singh N, Singh G, Singh HN, Rohilla R, Singh US (2003) The aromatic rices of Uttar Pradesh and Uttaranchal. In: Singh RK, Singh US (eds) A treaties on the scented rices of India. Kalyani, Ludhiana, pp 403–420

Thakur, Pran M (2010, May 13) Hindustan, Hindi Daily, Bhagalpur

Thakur R, Choudhary VK, Singh JP (2003) The aromatic rice of Bihar. In: Singh RK, Singh US (eds) A treaties on the scented rices of India. Kalyani, Ludhiana, pp 251–260

Tiwari, Amarendra K (2011, Sept 15) Dainik Jagaran, Hindi Daily, p 2

Part II
Nature and Problems of Agricultural Diversification

Chapter 9
Role of Dairying in Diversification of Indian Agriculture

C. L. Dadhich

The close relations of people and cattle benefited both species for 10000 years. They still do The Bin Luxor Egypt 1450 BC

9.1 Introduction

Livestock rearing in general and dairying in particular occupied an important place in Indian agriculture for centuries. However, milk production was mainly for home consumption till the advent of demand-led white revolution in the late 1960s. Demand for milk increased rapidly with urbanization in India. Largely, production of milk also kept pace with demand for milk. However, not all the states responded positively to demand. Majority of states in southern, western and northern regions responded positively, while eastern and north eastern states gave lukewarm response. As a result, spatial imbalance in milk production was further accentuated causing regional disparity in income generation in rural areas. The pace of diversification of agriculture was also adversely affected in these regions (Candler and Kumar 1998; Heredic 1999).

The primary objective of this chapter is to evaluate the growth pattern of milk production in India and suggest measures for upscaling milk production particularly in the north eastern and eastern region. Keeping in view the importance of dairying in states of these regions, four states viz., Assam, Bihar, Orissa and West Bengal were purposively selected for this analysis. The rest of the chapter is organised as follows. Section 9.2 gives brief account of Indian dairy economy and its role in diversification of agriculture; Sect. 9.3 deals with the present position of milk production in selected states. The potential for dairy development in these states is discussed in Sect. 9.4. Suggestions and conclusions are given in Sect. 9.5.

C. L. Dadhich (✉)
Indian Society of Agricultural Economics, Mumbai, Maharashtra, India
e-mail: isae@bom7.vsnl.net.in

M. Ghosh et al. (eds.), *Diversification of Agriculture in Eastern India*,
India Studies in Business and Economics, DOI 10.1007/978-81-322-1997-2_9

9.2 Dairy Economy

In 2009–2010, India produced 112.5 million t of milk constituting 15.8 % of the milk production of the world. India is the largest producer of milk in the world. The value of milk output for 2009–2010 was ₹ 228,809 crore that was higher than value of paddy (₹ 135,307 crore) and wheat (₹ 103,226 crore). The dairy sector employs about 10 million people in principal status and almost a million people as secondary occupation (Balakrishnan 2008). There are more than 70 million households engaged in milk production in India. India is among the most cost-effective producers of milk in the world (IFCN 2008). A unique feature of Indian dairy is the high share of buffalo milk at almost 53 %. Growth in milk production accelerated in last four decades coinciding with the implementation of operation flood programme of National Dairy Development Board (NDDB) (various years) envisaging dairy development through producer owned dairy co-operative structure. Over the last four decades, both population and food grain output grew at around 2 %, while milk production grew at about double of population. This facilitated a consistent and secular upward shift in per capita availability of milk from 112 g/day in 1970–1971 to 263 g/day in 2009–2010 (Patel 2004; Shah and Dave 2010; Shukla and Brahmankar 1999).

The contribution of agriculture gross domestic product (GDP) to total GDP has been falling substantially, from a level of 23 % in 1999–2000 to 14.6 % in 2009–2010 at constant prices. In contrast, decline of only one and half per cent is witnessed in the contribution of livestock GDP to total GDP during 1999–2000 to 2009–2010 at constant prices. This suggests that while the importance of agriculture is waning in total GDP, the significance of livestock is gaining importance in agricultural sector—the share of livestock in agriculture GDP increased from 23.1 % in 1999–2000 to 33.5 % in 2009–2010. Incidentally, dairy sector contributes more than two thirds to livestock GDP.

The stability in livestock income is stronger compared to the income deduced from agricultural activities. The analysis of national accounts statistics reveals that while the agricultural growth has been fluctuating over the years, the growth in value of milk has remained steady. This is a unique characteristic of the sector, and it is empirically found that the economic reliance of this sector has increased during periods of environmental adversities (Table 9.1).

In 2009–2010, the highest milk production was recorded in Uttar Pradesh (20.2 million t) distantly followed by Andhra Pradesh (10.4 million t), Rajasthan (9.5 million t), Punjab (9.3 million t), Gujarat (8.8 million t), Maharashtra (7.6 million t), Madhya Pradesh (7.1 million t), Bihar (6.1 million t), Haryana (6.0 million t), Tamil Nadu (5.7 million t), Karnataka (4.8 million t), West Bengal (4.3 million t), Kerala (2.5 million t), Orissa (1.7 million t), Jharkhand and Uttarakhand (1.4 million t each). These 16 major milk-producing states accounted for 95 % of milk production in the country. Three major milk-producing states of eastern region were Bihar, West Bengal and Orissa. Assam, the major milk producing state of north eastern region, was also selected for in-depth analysis (Fig. 9.1).

Table 9.1 Value of agricultural and milk output at constant price 2004–2005. (Source: Government of India 2010)

Item	2004–2005	2005–2006	2006–2007	2007–2008	2008–2009	2009–2010
Agriculture	458,495	484,588	503,122	532,555	525,706	520,663
Percentage variation over previous year	–	5.7	3.8	5.8	(–)1.3	(–)1.0
Milk	123,907	129,729	133,900	138,643	143,350	148,687
Percentage variation over previous year	–	4.6	3.2	3.5	3.4	3.7

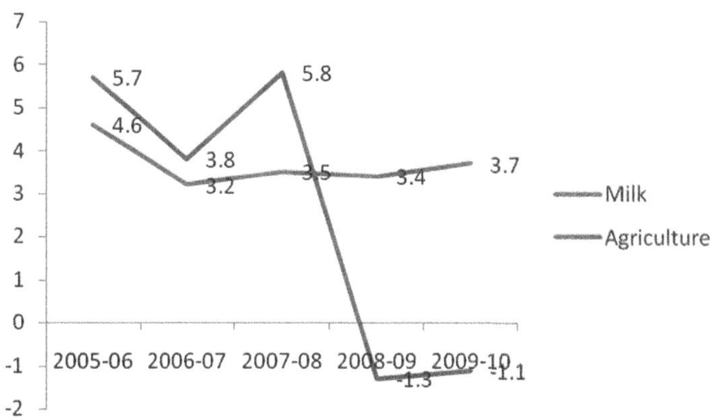

Fig. 9.1 Year on year variation in value of agriculture and milk output

9.3 Dairying in Selected States

In 2009–2010, among the selected states, highest milk production at 6.1 million t was recorded in Bihar (per capita milk availability of 175 g/day), followed by West Bengal 4.3 million t (per capita availability 133 g/day), Orissa 1.7 million t (per capita availability 112 g/day) and Assam 0.8 million t (per capita availability 69 g/day). Unlike the share of buffalo milk at 53 % for all India level, selected states have only 25 % buffalo milk. However, at disaggregate level, buffalo milk constituted 45 % in Bihar, 4 % in West Bengal and 12.5 % each in Orissa and Assam. Interestingly, this is a cow milk belt.

9.3.1 Low Proportion of Dairy Stock

As a result of demand-led white revolution coupled with farm mechanisation in the country, there was a shift in favour of dairy stock from work stock for the country

Table 9.2 State-wise proportion of livestock to dairy stock (in 000s). (Source: Government of India 2010)

States	Livestock	Of which dairy stock	% of dairy stock to livestock
Assam	17,227	5463	31.7
Bihar	30,342	13,160	43.3
Orissa	23,057	6443	27.9
West Bengal	37,419	12,445	33.2
All selected states	108,045	37,551	37.1
All India	529,668	201,196	37.9
Memo item Gujarat	23,513	15,522	66.0

Table 9.3 Statewise distribution of dairy stock in 2007 (percentage to total). (Note: Figures in brackets are absolute, Source: Government of India 2010)

States	Type of stock			Total
	Crossbred cow	Indigenous cow	Buffalo	
Assam	5.3	89.6	5.11	100 (5463)
Bihar	12.4	46.6	41.0	100 (13,160)
Orissa	12.8	78.0	9.2	100 (6443)
West Bengal	16.7	81.1	2.2	100 (12,465)
All selected states	12.8	69.7	17.5	100 (37,531)
All India	13.0	44.4	42.6	100 (201,196)

as a whole. However, data presented in Table 9.2 reveal that barring Bihar, the proportion of dairy stock to total stock was lower in the selected states as compared to all India level. Lowest proportion of about 28 % was noticed in the state of Orissa. Incidentally, in advanced dairy state, this proportion was as high as 66 %. This suggests that demand-led dairy development was at low ebb in these states.

9.3.2 *Preponderance of Low-Yielding Dairy Stock*

The composition of dairy stock as presented in Table 9.3 reveals that Bihar had the highest dairy stock numbering (13.1 million) followed by West Bengal (12.4 million), Orissa (6.4 million) and Assam (5.5 million). The proportion of buffalo was highest at 41 % in Bihar distantly followed by Orissa at 9.2 %, Assam at 5.1 % and West Bengal at 2.2 % as against 42.6 % at all India. The proportion of indigenous cow ranged from 89.6 % in Assam to 46.6 % in Bihar with an average of 69.7 % for the region. All India average was 44.4 %. However, the proportion of crossbred cow ranged between 5.3 % in Assam and 16.7 % in West Bengal with average of 12.8 % for the region. All India average was higher at 13.0 %. It is indeed misleading to compare the status of crossbreeding programme on the basis of total dairy stock. It should be compared as a proportion of total female cattle. Taking this point in view, crossbred cattle as a proportion of total cattle were lower at 15.5 % for the region as compared to 21.7 % for all India. Incidentally, proportion of crossbred cow was as high as 83.6 % for Punjab (Table 9.4). However, taking together buffalo and

Table 9.4 Status of crossbreeding programme of cattle in selected states 2007. (Note: Figures in brackets are percentage to total, Source: Government of India 2010)

States	Number of cattle (in 000)		
	Crossbred	Indigenous	Total
Assam	288 (5.5)	4895 (94.5)	5183 (100.0)
Bihar	1632 (21.1)	6127 (78.9)	7759 (100.0)
Orissa	823 (14.1)	5029 (85.9)	5852 (100.0)
West Bengal	2084 (17.1)	10,114 (82.9)	12,198 (100.0)
All selected states	4827 (15.5)	26,165 (84.5)	30,992 (100.0)
All India	24,686 (21.7)	89,325 (78.3)	114,011 (100.0)
Memo item Punjab	1077 (83.6)	212 (6.4)	1289 (100.0)

Table 9.5 Average milk yield in selected eastern states 2009–2010 (kg/day). (Source: Government of India 2010)

State	Category of in milk animal			
	Crossbred cow	Indigenous cow	Buffalo	Average
Assam	3.34	0.95	2.5	1.27
Bihar	6.19	2.91	3.92	3.72
Orissa	5.94	1.17	2.90	2.06
West Bengal	5.98	1.83	4.59	2.76
All selected eastern states	5.80	1.81	3.76	2.78
All India	6.87	2.14	4.57	3.94
Highest average state yield (Punjab)	10.41	5.64	8.51	8.88

crossbred cow, which are considered as high-yielding stock, the proportion was 30% in the region as compared to 56% for all India. Thus, the selected states have preponderance of low-yielding stocks.

9.3.3 *Poor Yield of Dairy Stock*

It may be observed from the data presented in Table 9.5 that the yield of indigenous cattle was as low as 0.9 kg/day in Assam, 1.2 kg/day in Orissa and 1.8 kg/day in West Bengal as against 2.1 kg/day for all India. Incidentally, the highest yield of 5.64 kg/day per animal was noticed in Punjab. The higher proportion of indigenous cattle with poor yield level have plagued dairy development in this region. However, except Assam, the yield level of buffalo and crossbred cattle was almost close to that of all India average but far below the highest yield noticed in the state of Punjab.

9.3.3.1 Skewed Distribution of Milk Production

Because of high proportion of low-yielding animals in this region, large number of animals has produced small quantity of milk (Table 9.6). Overall, this region

Table 9.6 State-wise dairy stock and milk production 2007 (in %). (Source: Government of India 2010)

State	Share of state dairy stock to total dairy stock	Share of state milk productivity to total milk production
Assam	2.7	0.7
Bihar	6.5	5.5
Orissa	3.2	1.5
West Bengal	6.2	3.9
All selected states	18.6	11.6
Remaining states	81.4	88.4
All	100.0	100.0

accounted for 18.6 % of total dairy stock in the country contributing only 11.6 % of milk. This indicates that the remaining 81.4 % dairy stock produced 88.4 % of milk in other states. State-wise, at the disaggregate level, Assam had 2.7 % dairy stock producing only 0.7 % of the total milk in the country, followed by Orissa with 3.2 % dairy stock producing 1.5 % milk, West Bengal with 6.2 % dairy stock producing 3.9 % milk and Bihar with 6.5 % dairy stock producing 5.5 % of milk produced in the country. In short, the distribution of dairy stock and milk production is highly skewed in the region.

9.3.3.2 Poor Dairy Infrastructure

Dairy infrastructure consists, among others, installed milk processing capacity, artificial insemination performed, number of village dairy co-operative societies and membership thereof. In the absence of number of villages covered and pouring members, milk procured by dairy co-operatives has been used as proxy for the purpose of this analysis. It may be observed from data presented in Table 9.7 that about 40 artificial insemination were performed for every 100 milch at all India level. In the eastern region, more than 30 artificial inseminations were performed for every 100 milch, which is far below the average of the nation. In Assam and Bihar the performance was inadequate, while it was above average in West Bengal and closer to average in Orissa. Overall, artificial insemination service was far from satisfactory. Similarly, installed processing capacity in the region was inadequate. As against 8 % procurement of milk production at the national level, this region has achieved only 2.4 % while the number of dairy co-operatives organised in the region compared well with milk produced in the region (as against 11.6 % milk produced the proportion of the Societies organised was 10 %), membership per society was strikingly low as compared to the national level. Similarly, no especial efforts have been made in the region to augment the fodder supply through cultivation of fodder crops. The area under fodder crops is not worth reporting. Overall, inadequate dairy infrastructure is equally responsible for poor participation of the region in dairy development.

Table 9.7 Dairy infrastructure 2009–2010. (Source: Government of India 2010)

State	Milch stock (in 000)	Artificial Insemination performed (in 000)	Percentage of Col 3 to Col 2	Milk production kg/day (in 000 kg)	Milk procured by Co-ops. Kg/day	% col 6 to col 5	Processing capacity installed kg/day (in 000)	% of col 8 to col 5
Assam	2852	204	7.2	2071	5	0.2	–	–
Bihar	4672	950	20.3	16,800	700	4	785	4.7
Orissa	2990	1166	39	4500	200	5	199	4.4
West Bengal	6233	2776	44.5	11,800	300	2	2081	17.6
All selected states	16,747	5096	30.4	35,171	1205	3.4	3065	8.7
All India	111,090	44,621	40.2	319,000	25,900	8	98,316	30.8

9.4 Huge Growth Potential

In this dark cloud, there is of course, a silver line in the form of satisfactory average yield of crossbred cattle and buffalo. Apparently, barring Assam, milk producers in the region gave adequate attention to high-yielding animals. Thus, with the support of improved technology (i.e. proper breeding and feeding), dairying should be made a viable and high-paying proposition. There is enough scope for upgrading the genetics of cattle, by crossbreeding with native yielding breeds if not with the exotic breeds. Availability of permanent pasture and other grazing land particularly in Assam and Orissa is another favourable feature of the region. Low per capita availability of milk and growing demand of milk will sooner than later force the producers to adopt improved technology envisage under the National Dairy Plan of NDDB. Bihar is already on the growth path, other states should follow the suit. Keeping in view the growth potential of dairy sectors, the 11th Five Year Plan envisages a growth of 6 to 7% per annum for dairy and allied sectors to achieve a stipulated growth of 4%.

9.5 Suggestions and Conclusions

While milk yield of indigenous cows was striking, low yield of crossbred cattle was almost close to the average yield at the national level. This suggests that crossbreeding programme of cattle is technically feasible in the eastern region. Like the first green revolution, first white revolution has also bypassed this region. Market-led first white revolution centred on comparatively better milk yield areas like Punjab, Haryana, Gujarat, Karnataka etc. This resulted in shifting of milch animals to these areas, and higher production of milk was through increase in herd size rather than enhancing the productivity of milch animals. Concerted efforts are needed to take up crossbreeding of existing low-yield indigenous stock with exotic or improved native breeds. In this context, it is important to note that some of the upgraded Indian breeds of cattle viz., Gir, Red Sindhi Sahiwal etc. are popular in other parts of the world. The slow growth of milk production in the major milk producing states made it increasingly important to usher in technology-based and market-led second white revolution in this region along with second green revolution as envisaged in the 11th Five Year Plan. The higher milk production in the region will go a long way in diversification of agriculture. It will not only ensure higher rural income but also nutritional security in the region that has been very low as compared to per capita availability of milk in the country. Efforts are also required to improve the wet ratio of existing indigenous stock through proper feeding and management of animals.

Diversification of agriculture is imperative to ensure, among others, nutritional security and smooth flow of rural income stream. While value of agricultural output was highly volatile, value of livestock output in general and value of milk output in particular were steady and stable. Dairy sector is one of the main drivers of growth of Indian agriculture. Its importance cannot be overemphasised in the context of diversification of agriculture. Demand-led white revolution made rapid strides in majority of Indian states but largely by passed eastern region, while most of the states registered shift in the composition of livestock in favour of dairy stock, the eastern states did not witness perceptible shift in composition of livestock. The study brings to the fore that preponderance of low-yielding indigenous cattle in dairy herd has adversely affected the participation of this region in white revolution. Besides, inadequate infrastructure has also caused damage to the growth of dairy sector and consequently non-diversification of agriculture. However, limited yet satisfactory induction of crossbreeding programme of cattle in the region indicates huge growth potential for dairy development. It does without saying that upgradation of indigenous cattle if not the exotic crossbreeding programme in a big way will place dairy sector on a fast growth trajectory and go a long way in diversification of agriculture in the region.

References

Balakrishnan R (2008) Supply chain management for Indian agriculture. Indian Merchants' Chamber and IMC Economic Research and Training Foundation, Mumbai

Candler W, Kumar N (1998) India: the dairy revolution. World Bank, Washington, D.C.

Government of India (2010) Basic animal husbandry statistics 2010. Department of Animal Husbandry, Dairying & Fisheries, Ministry of Agriculture, Government of India, New Delhi

Heredia R (1999) The AMUL India story. Tata McGraw Hill, Noida

IFCN (2008) IFCN dairy report. International Farm Comparison Network (IFCN), Kiel, Germany

National Dairy Development Board (2000–2010) Annual report of various years

Patel A (2004) Ensuring a future for cooperatives in India: the law and other challenges. Centre for Management Development, Thiruvananthapuram

Shah J, Dave D (2010) A shift from crop-mixed traditional dairying to market-oriented organised dairy farming—plausible factors responsible for structural transformation in Indian dairy sector. Indian J Agric Econ 65(2):298–307

Shukla RK, Brahmankar SD (1999) Impact evaluation of operation flood on rural dairy sector. National Council of Applied Economic Research (NCAER), New Delhi

Chapter 10
Crop Diversification Through Oilseeds in Eastern India

M. K. Bhowmick, B. Duary and P. K. Biswas

10.1 Introduction

Among the countries in global oilseed economy, India ranks fourth, next to USA, China and Brazil in terms of vegetable oils (Rao 2008). In India, oilseeds represent the second largest agricultural commodity after food grains in terms of area, production and value (Reddy and Suresh 2009). However, the contribution of the country to the global oilseed production is still very low, which is mainly attributed to the extremely low productivity of different oilseed crops. This could be due to the fact that energy-rich crops are generally grown under energy-starved conditions. In fact, most of the oilseeds are raised on marginal and sub-marginal lands unlike cereals grown mostly under fertile soils with assured irrigation facilities. Besides, nearly 76 % of oilseeds area in the country is rainfed, which is often subjected to vagaries of monsoon (Hegde 2005b). This sort of disparity along with uneven distribution of irrigated area has made oilseed cultivation less productive and risky. Though the country made a record production of oilseeds (27.98 million t) during 2005–2006, it had to import a substantial amount of vegetable oils, thereby draining out a huge amount of its foreign exchange for purchasing vegetable oils in the world market. In this backdrop, there is a formidable challenge to increase oilseed production in order to feed the burgeoning population along with the maintenance of soil health in the long run (Bhowmick et al. 2006), especially in eastern India occupying about

M. K. Bhowmick (✉)
Rice Research Station (Government of West Bengal), Chinsurah (R.S.), Hooghly 712102, West Bengal, India
e-mail: bhowmick_malay@rediffmail.com

B. Duary · P. K. Biswas
Institute of Agriculture (Palli Siksha Bhavana), Visva-Bharati, Sriniketan, Birbhum 731236, West Bengal, India
e-mail: bduary@yahoo.co.in

P. K. Biswas
e-mail: ppabitra07@rediffmail.com

M. Ghosh et al. (eds.), *Diversification of Agriculture in Eastern India*,
India Studies in Business and Economics, DOI 10.1007/978-81-322-1997-2_10

28 % of country's total geographical area (Zaman 2012). The eastern region of the country is most backward in terms of per capita income, agricultural growth and infrastructure development. The yield levels are low because of the uncertain production environment and poor adoption of improved varieties and technologies. Overall, the region was food-based concentrating largely on rice, with little diversification. The humid atmosphere and high rainfall makes cultivation of rice more favourable in this region, which, however, has also gradually been emerging as an oilseed producing region with cultivation of crops like rapeseed-mustard, groundnut, sesame and soybean (Joshi 2005).

Considering economic viability, food and nutritional security, and long-term sustainability, particularly in the fast changing global scenario, it is, therefore, essential to achieve self-sufficiency in oilseed production of the country. This warrants immediate corrective steps to be taken towards diversification through oilseeds in traditional crops and cropping systems.

10.2 Conceptualizing Crop Diversification

Diversification is defined as diverting a sizeable acreage from the existing crop system to some other alternate crops, cropping systems or farm enterprises while maintaining a general equilibrium of meeting the needs for food, fodder, fibre and fuel, and while simultaneously taking care of basic soil health and productivity of agro-ecosystem of the area at large (Kalra 1990; Gautam and Sharma 2004). Thus, it is a strategy of shifting from less profitable to more profitable crops; changing of varieties and cropping systems; increasing exports and competitiveness in both domestic and international markets; protecting the environment; and making conditions favourable for combining different enterprises (Gautam and Sharma 2004).

Crop diversification is one of the major components of diversification in agriculture. It is an important tool for achieving the objectives of food security, nutrition security, income growth, poverty alleviation, employment generation, judicious use of natural resources, sustainable agricultural development and environmental and ecological improvement (Gautam and Sharma 2004; Sarkar 2005). The current thinking of diversification in the contemporary agriculture has, therefore, arisen so as to make it sustainable and profitable besides adding insurance against the disease outbreaks/insect epidemics/weed resurgence and socio-economic advantages. In addition, it is important to ensure that crop intensification goes together with crop diversification. If intensification enhances the productivity, diversification adds strength to the farmer in terms of sustainability (Siddiq 1999; Ali and Kumar 2002; Ali 2004). However, crop diversification in India is generally viewed as a shift from traditionally grown less remunerative crops to more remunerative crops (Reddy and Suresh 2009).

10.3 Need for Diversification in Cereals and Cereal-Based Cropping Systems

Among the cereal crops, rice is the most important and staple food crop for more than two-thirds of the Indians (Mishra 2005). It is cultivated in 44.8 million hectares (m ha) area in the country, of which eastern India, comprising Assam, West Bengal, Orissa, Jharkhand, eastern Uttar Pradesh and Chhatisgarh, accounts for 27.8 m ha (62 % of total rice area), contributing only 49 % of the total rice production. Based on the hydrology and topography of the land, rice area is divided into different ecologies: rainfed (uplands and lowlands), irrigated and hill rice. In all, 73 % of total rice growing area in eastern India is rainfed (20 % in upland and 53 % in lowland). The rainfed upland agro-ecosystem comprises of the soils poor in fertility and with low moisture retention capacity. The net economic return from this ecosystem also fluctuates from year to year due to vagaries of the southwest monsoon and traditional cropping system, dominated by rice. In such a situation, crop diversification and rice substitution with comparatively low water requiring crops like oilseeds may prove as an effective means to improve sustainability and rain water utilization efficiency, and in turn, productivity of watershed (Mahapatra 2002). The rainfed lowlands, where nutrient status and moisture availability is enough, may support, if managed properly, short-duration oilseeds for double cropping (Ali and Mishra 2002). On the other hand, rainfall being erratic and uncertain, expansion of oilseed area under irrigation is crucial for effective changes in cropping pattern, and adoption of profitable crops and technologies in the production systems to minimize risks considerably (Rai 2002).

Unfortunately, most of the high productivity systems are cereal-based, having high resource demand and continuously practiced in major parts of the country (i.e. rice-wheat in Indo-Gangetic plains (IGPs), rice-rice in coastal and high rainfall areas, and coarse-cereal-based systems in low rainfall areas) over the decades (Gangwar and Prasad 2005). Sustainability of these exhaustive systems is threatened owing to the emerging second generation problems viz. decline in factor productivity; depletion of soil fertility due to over-mining of native nutrient reserve; decline in fertilizer use efficiency and soil organic matter content; emergence of macro- and micro-nutrient deficiencies; lowering or rising of water table; water quality deterioration in coastal areas; soil degradation; increasing problems of salinity–alkalinity; resurgence of diseases, insects, and weeds; environmental pollution/degradation; formation of hardpan; stagnant farm incomes, etc. (Siddiq 1999; Ali and Kumar 2002; Gautam and Sharma 2004; Gangwar and Prasad 2005; Reddy and Suresh 2009). Crop diversification shows lots of promise in alleviating these problems through fulfilling the basic needs and regulating farm income, withstanding weather aberrations, controlling price fluctuation, ensuring balanced food supply, conserving natural resources, reducing the chemical fertilizer and pesticide loads, environmental safety and creating employment opportunity (Gill and Ahlawat 2006).

Today, the increasing demand for oilseeds to meet the requirement of ever-increasing population has brought the non-traditional and unutilized areas in sharp focus, as these crops can thrive well under such circumstances where other crops do not grow well (Ali and Mishra 2002).

Among all the states in the country, West Bengal continues to occupy the first position in rice production while the state has the second position in potato production and again the foremost position in vegetable production (Anonymous 2005). Even after this increase in the production of rice, potato and vegetables, there are still deficits in the production of certain other food grains and crops. The wheat production during 2009–2010 was 8.47 lakh t which is less than the present requirement of about 20 lakh t. Likewise, the production of pulses during 2009–2010 was 1.50 lakh t and of oilseeds 7.26 lakh t, which were again less than the present requirement of the state (Anonymous 2011). Therefore, some steps need to be taken for increasing production of the said important crops so as to reduce the deficits in such cases, and to ensure remunerative prices to the farmers because of excess demand for these crops.

All these call for crop diversification including partial substitution with oilseeds as well as adoption of other enterprises under specific situations for increasing oilseed production not only in the state of West Bengal but also in the eastern India as a whole. In addition, there is a need to develop suitable production technology of these alternative farming systems to make them equally or even more competitive with the existing systems. Besides, it is essential to use various resources rationally for sustained productivity and conserve biodiversity for achieving food and nutritional security (Bhowmick et al. 2006).

10.4 Options for Diversification and Area Expansion with Oilseeds

In respect of growth in oilseed production in India, nearly 55 % of the increase was achieved by area expansion and 45 % by productivity improvement (Hegde 2005a). Area expansion has occurred in favour of those oilseed crops, which have shown a higher growth rate of productivity due to technological development (e.g. rapeseed-mustard) or whose relative prices with competing crops have moved in their favour (e.g. sunflower) or whose total profitability has increased due to a combination of technological development and higher prices as in soybean (Hegde 2000a, 2000b, 2005a, 2005b). There is hardly any scope to bring additional area exclusively under oilseeds due to the increasing demand for all crops and commodities. Introduction of high-yielding varieties of oilseed crops (Table 10.1) can replace a number of traditional low-yielding crops because of their higher efficiency in utilizing rainfall and moisture, and ultimately, resulting in higher yields and returns (Reddy and Suresh 2009). Therefore, there is a need to find newer niches and situations where oilseeds can be introduced. The possible options available for crop diversification with oilseeds by going in for their area expansion (Hegde 2000a, 2002, 2004, 2005a, 2005b; Palaniappan and Jeyabal 2002; Sengupta and Das 2003) are as follows:

Table 10.1 Annual oilseed crops and their promising varieties/hybrids in West Bengal. (Source: Bhowmick et al. 2006)

Common name	Botanical name	Varieties/hybrids	Maturity (days)	Oil content (%)	Seed yield (q ha^{-1})
Groundnut	*Arachis hypogaea* L.	Varieties			
		AK 12-24	110–120	48.0–48.5	20–22 (Pod)
		Phule Pragati (JL 24)	100–110	50.7	20–25 (Pod)
		GG 2	110–120	49.0	20–25 (Pod)
		VRI 4	115–125	49.0	22–25 (Pod)
		ICGS 44	110–120	49.0	22–25 (Pod)
		J 11	115–120	49.0	18–20 (Pod)
Toria (Rapeseed)	*Brassica rapa* L. var. *Toria*	Agrani (B 54)	70–75	40.0	8–10 (R)
		Panchali (TWC 3)	80–85	40.0	10–12 (R) and 12–14 (I)
		T 9	90–95	40.0	12–15 (R&I)
		PT 303	95	43.0–44.0	9–11 (I)
Yellow sarson/ Swet sarson/ Colza (Rapeseed)	*Brassica rapa* L. var. yellow sarson	Binoy (B 9)	90–95	46.0	14–15 (I)
		Subinoy (YSB 19-7C)	95–98	46.0	14–16 (I)
		Jhumka (NC 1)	95–100	45.0	15–17 (I)
		NRCYS 05-02	113	42.9	14–15
		Narendra Sarson 2 (NDYS 2)	125–130	43.0	11–12
Gobhi sarson (Rapeseed)	*Brassica napus* L.	Kalyan (WBBN 1)	105–110	43.0	12–15 (I)
Rai/Indian mustard/ Brown mustard (Mustard)	*Brassica juncea* (L) Czern. & Cosson.	Seeta (B 85)	95–100	38.0	12–14 (R)
		Sarama (RW 85-89)	110–115	38.0	17–19 (I)
		Bhagirathi (RW 351)	110–115	38.0	16–18 (I)
		Sanjukta-Asech (RW 4C-6-3/II)	95–100	38.0	10–12 (I)
		Varuna (T 59)	120–130	43.0	20–22 (I)
		Pusa Bold (PR 45)	110–120	42.0	18–20 (I)
		Krishna (PR 18)	132	40.0	13–14 (I)
		Kranti (PR 15)	125–130	40.0	15–18 (I)
		RH 30	130–135	39.0	16–20 (R&I)
		JD 6	81–114	39–44	6–10
		Vardan (RK 1467)	120–125	40.0	10–16 (I)
Soybean	*Glycine max* (L.) Merr.	Bragg	112–115	22.4	15–20
		JS 80-21	105–110	–	25–30
		JS 2	90–95	–	18–20
		JS 335	95–100	–	25–30
		Birsa Soy 1	100–110	–	20–24
		Indira Soy 9	106	–	22–23
		MAUS 71	93–100	–	18–30

Table 10.1 (continued)

Common name	Botanical name	Varieties/hybrids	Maturity (days)	Oil content (%)	Seed yield (q ha^{-1})
		PK 472	100–106	–	25–30
		PK 564	105–115	–	30–35
		Pusa 16	105–115	–	25–30
		Pusa 22	110–120	–	25–30
		Pusa 24	110–115	–	25–30
		Pusa 37	110–115	–	25–30
		Shilajeet	100–105	–	20–25
		RAUS 5	96–104	–	30–35
Sunflower	*Helianthus annuus* L.	Hybrids			
		KBSH 1	90–95	42.0–44.0	12–16 (R) and 16–24 (I)
		KBSH 44	90–100	36.0–40.0	12–16 (R) and 16–25 (I)
		MSFH 17	90–95	36.0–38-0	13–16 (R) and 16–25 (I)
		DRSH 1 (PCSH 243)	95–105	40.0	13–16
		Jwalamukhi(PSCL 5015)	95–100	40.0–42.0	12–16 (R) and 16–24 (I)
		PAC 36	95–100	40.0–42.0	12–16 (R) and 16–24 (I)
		PAC 1091	95–100	40.0–43.0	12–16 (R) and 16–24 (I)
		Sunbred 212	109	–	15
		Sunbred 275	120	42.0	15–16
		MSFH 8	90–95	42.0–44.0	12–15 (R) and 15–18 (I)
		BSH 1	85–90	40.0–42.0	8–14 (R) and 15–18 (I)
		Pro Sun 09	90–95	42.0–44.0	12–16 (R) and 16–25 (I)
		MLSFH 47	92–98	40.0–41.0	12–16 (R) and 16–24 (I)
		Surya	90–100	38.0–40.0	8–10 (R) and 10–12 (I)
		Varieties			
		Morden (Cernianka 66)	80–90	38.0–40.0	6–7 (R) and 10–12 (I)
		Sidheshwar (LS 11)	85–88	40.0	10 (R)
		EC 68414 (Peredovik)	100–115	40.0–44.0	6–8 (R) and 12–15 (I)
		DRSF 108	95–100	37.0	12–14
		DRSF 113	90–98	40.0	12–14
		Morden (Cernianka 66)	80–90	38.0–40.0	6–7 (R) and 10–12 (I)

Table 10.1 (continued)

Common name	Botanical name	Varieties/hybrids	Maturity (days)	Oil content (%)	Seed yield (q ha^{-1})
Sesame	*Sesamum indicum* L.	Tilottama (B 67)	75–80	40.0	6–8 (R) and 8–10 (I)
		Rama (Improved Selection 5)	80–85	45.0	8–10 (R) and 10–15 (I)
		Krishna (RAUSS 17-4)	85–90	45.8–46.0	8–10 (R) and 10–11 (I)
		HT 1	65–70	50.0	5–6 (Post-*kharif*)
		PT 1	65–70	50.0	5–6 (Post-*kharif*)
		Uma (OMT 11-6-3)	80–85	51.0	9–11
		Savitri (SWB 32-10-1)	84–87	51.1	8–10
Safflower	*Carthamus tinctorius* L.	Hybrids			
		DSH 129	120–125	30.0	18–22
		MKH 11	130–135	31.0	17–18
		NARI NH-1(PH 6)	135–140	31.0	18–19
		Varieties			
		BLY 652	120–130	38.0	9–10
		Annigeri 1 (A 1)	115–125	29.0–35.0	9–10
		APRR 1	110–120	35.0	10–12
		Malvia Kusum (HUS 305)	165	36.0	15
		NARI 6	137	30.0	15
Niger	*Guizotia abyssinica* (L.f.) Cass.	Shyadri (IGP 76)	100–105	40.0	4.5–5.0
		No.71	95	42.0	4.5–5.0
		Bhawani (GA 5)	120–125	39.0	4.0
		Shiva (GA 10)	115–120	42.0	4.0–5.0
Linseed	*Linum usitatissimum* L.	Neela (B 67)	120–125	40.0	7–9 (R) and 9–10(I)
		Garima	125–130	42.0	12–14 (I)
		Shubhra	130–135	45.0	9–11 (R) and 11–13(I)
		Gaurav (LCK 152)	135–140	43.0	9.5 (fibre) and 10.5 (seed) (I)
Castor	*Ricinus communis* L.	Hybrids			
		GCH 4	150–180	50.0–51.0	11–13 (R) and 20–22 (I)
		GCH 5	180–240	52.0–57.0	18 (R) and 25 (I)
		GCH 6	210	48.09	30 (I)
		DCH 177	90–100	49.0	15 (R) and 20 (I)
		DCH 32	85–95	48.0	18 (R) and 28 (I)

Table 10.1 (continued)

Common name	Botanical name	Varieties/hybrids	Maturity (days)	Oil content (%)	Seed yield (q ha^{-1})
		DCH 519	105–110	50.0	17–21
		Varieties			
		B 1	120–130	45.3	3–5
		Jyoti (DCS 9)	90–150	49.0	10 (R)
		Kranti (PCS 4)	90–150	48.0–50.0	13–14 (R)
		Jwala (48-1)	130–220	48.0	10 (R) and 18 (I)
		Aruna	120–150	52.0	10 (R) and 20 (I)

I Irrigated *R* Rainfed

1. Extending cultivation in underutilized farming situations such as rice-fallows of eastern India where more than 15 m ha are under low land rice;
2. Intercropping of oilseeds in nearly about 45 m ha of widely spaced crops like sugarcane, maize, cotton, pigeon pea and plantation crops, and in less remunerative traditional staple food crops where replacement is not possible;
3. Substitution of existing low-yielding and high water requiring crops (*boro* rice, wheat, sugarcane, etc.) with high-yielding and less water intensive oilseed crops (sunflower, *Brassica*, sesame or groundnut) in tail-end of canals and in areas irrigated by wells and tanks where water availability is limited, and also in saline areas;
4. Cultivation of oilseeds in newer seasons e.g. expanding areas particularly under *rabi* groundnut cultivation as this crop registers high productivity during *rabi;*
5. Introduction of suitable oilseed crops through diversification of rice–rice and rice–wheat cropping systems;
6. Growing oilseeds (sunflower, sesame or niger) under contingency planning where the season for regular crops is not conducive or when these have failed;
7. Fitting the oilseeds (sunflower, sesame, etc.) as catch crops in the period left between two regular crops;
8. Value addition to some of the main and by-products of oilseeds which will increase their profitability, help expand the area and also arrest constant decline in area observed recently in linseed, safflower, etc.;
9. Growing of minor, non-conventional or tree-borne oilseeds in the forest areas, waste lands and marginal lands.

The diverse agro-ecological conditions in the country are favourable for growing nine annual oilseeds (Table 10.1), including seven edible sources (groundnut, rapeseed-mustard, soybean, sunflower, sesame, safflower and niger) and two non-edible sources (castor and linseed) apart from a number of minor oilseeds (Table 10.2) and perennial oilseeds (Table 10.3) of horticultural and forest origin, including coconut and oil palm in particular (Rai 2002; Hegde 2000a, 2005a; Reddy and Suresh 2009).

Table 10.2 Minor oilseeds of potential value and their distribution in India. (Sources: Rai 2002; Saxena et al. 2002)

Common name	Botanical name	Oil content (%)	Major producing states
Sal	*Shorea robusta*	12.5	Madhya Pradesh, Orissa and Bihar
Neem	*Azadirachta indica*	20.0	Andhra Pradesh, Gujarat, Madhya Pradesh, Tamil Nadu Maharashtra, Karnataka and Uttar Pradesh
Karanja	*Pongamia glabra*	30.0	Andhra Pradesh, Tamil Nadu and Karnataka
Mahua	*Madhuka longifolia*	35.0	Uttar Pradesh, Madhya Pradesh, Andhra Pradesh and Gujarat
Kusum	*Schleichera trijuga*	33.0	Bihar and Orissa
Kokum	*Garcinia indica*	40.0	Maharashtra and Goa
Rubber seed	*Hevea brasiliensis*	18.5	Kerala
Pinnai/Undi	*Calophyllum indophyllum*	60.0	Kerala
Khakan/Pilu	*Salvadora oboides*	33.0	Maharashtra and Gujarat
Mango kernel	*Mangifera indica*	8.0	Andhra Pradesh, Uttar Pradesh and Maharashtra
Tung	*Aleurities fordii*	–	Himachal Pradesh
Pisa	*Actinodaphne hookeri*	48.0	Karnataka and Maharashtra
Maroti/Kavathi	*Hydnocarpus wightianai*	33.0	Kerala
Nahor	*Mesua ferrea*	40.0	Tamil Nadu, Kerala, Andaman and Assam
Dhupa	*Vaterie indica*	17.0	Karanataka and Kerala
Tumba	*Citrullus colocynthis*	30.5	Rajasthan

Table 10.3 Perennial oilseeds with their oil yield potentials in India. (Source: Menon 1985)

Common name	Botanical name	% oil (Dry wt. basis)	Oil yield (t ha^{-1})
Coconut	*Cocos nucifera*	60 (kernel)	0.6–1.5
Oil palm	*Elacis guineensis*	47 (pulp) and 50 (kernel)	1.7–3.0
Macauba	*Acrochomia sclerocarpa*	61 (pulp) and 65 (kernel)	4.0–6.0
Tucuma	*Astrocaryum tucuma*	35–48 (pulp) and 45–48 (kernel)	1.5–2.0
Pinhao/Verendah	*Jatropha curcus*	60–65 (kernel)	1.5–2.3
Piqui	*Caryocar brasilience*	36.5 (kernel) and 6 (pulp)	2.3–3.5

The area expansion under oilseeds can further be promoted through ecological zoning. Delineating efficient zones for each oilseed crop helps realize potential yields with limited efforts and high input use efficiency. A serious effort must, therefore, be attempted to extend area where high oilseed productivity can be achieved (Sengupta and Das 2003). However, some potential regions for area expansion under different oilseed crops in eastern India have been presented in Table 10.4.

Table 10.4 Diversification of traditional crop base with annual oilseed crops in India. (Sources: Palaniappan and Jeyabal 2002; Hegde 2005a; Reddy and Suresh 2009)

Crop	Area suggested for diversification
Groundnut	As a replacement crop for minor millets in Bihar and Orissa
	As an intercrop with upland rice in north-eastern region and non-traditional areas in Uttar Pradesh (Bundelkhand, western Uttar Pradesh and *Tarai-bhabar* tract) and as a replacement crop of upland rice in Orissa, Tamil Nadu, Bihar and Andhra Pradesh during *kharif*
	As a *rabi*/summer crop on residual moisture of rice-fallows in Assam, West Bengal, Orissa and Chhatisgarh and as an irrigated crop in *Kosi* and *Tawa* commands in rice-fallows in Bihar and Madhya Pradesh
	As a second crop to substitute rice-rice with rice-groundnut system to prevent the build-up of insects and diseases in endemic areas where rice-rice system is prevalent
Rapeseed and mustard	As an introduction in *diara* tracts of northern and eastern India and Gujarat
	Mustard as an intercrop with autumn planted sugarcane, potato, wheat, lentil, chickpea, etc.
	As a rainfed *utera* crop in lowland rice-fallows in north-eastern states
	As a crop grown with brush fencing of fields to prevent cattles in upland rice-fallows in north-eastern hill states
Soybean	As a replacement crop for minor millets in Bihar and Orissa
	As an intercrop with maize, pigeonpea, pearlmillet, etc. in Madhya Pradesh, Rajasthan, Uttar Pradesh, Andhra Pradesh, Karnataka, Orissa and Tamil Nadu, and as a rotational crop in pest endemic areas of rainfed cotton
	As an introduction in non-traditional areas of north-eastern hills which are under agri-pastoral/ agri-silvicultural system
	As an intercrop with upland rice in north-eastern region and as a replacement crop for upland rice in Orissa, Tamil Nadu and Uttar Pradesh
	As a *rabi*/summer crop in rice-fallows in Assam and southern states
Sunflower	As a replacement crop of upland rice in Orissa, Tamil Nadu, Bihar and Andhra Pradesh
	As a crop in rice-fallows in most parts of rice growing areas
Sesame	As a summer crop under assured moisture condition after potato, rice etc. in eastern states and under limited water availability situation (three to four irrigations and shorter duration only) in central, peninsular and eastern India
Safflower	As a crop under limited moisture condition in traditional (Maharashtra, Karnataka and Andhra Pradesh) and non-traditional (Malwa plateau of Madhya Pradesh, Chhatisgarh, Udaipur region of Rajasthan and Bundelkhand region of Uttar Pradesh) areas
	As an intercrop/border crop/replacement sole crop for sorghum, coriander, dryland wheat, chickpea, linseed, etc. in the above states
	As an intercrop/border crop/replacement sole crop for sorghum, coriander, dryland wheat, chickpea, linseed, etc. in the above states
	As a replacement crop for desi cotton, minor millets and low-yielding pulses in the above states
Castor	As bund crop in all regions
	For replacing traditional varieties with high-yielding ones in Uttar Pradesh and Bihar

Table 10.4 (continued)

Crop	Area suggested for diversification
Linseed	As an intercrop of wheat, chickpea, lentil, coriander, etc. in linseed-growing areas
	As a *utera* crop in rice-fallows in eastern and north-eastern states
Niger	As a replacement sole crop/mixed crop for bajra, ragi, castor, groundnut and pulses under rainfed condition
	As an introduction in non-traditional areas under residual soil moisture and rainfed conditions

10.5 Strategies for Increasing Oilseeds Production

The state of West Bengal has much deficit in the production of oilseeds. Though the state has made remarkable progress in oilseed production in last few years (Table 10.5), it did not keep pace with the growing population as the estimated population of the state has risen from 802.21 lakhs (as per Census, 2001) to 913.48 lakhs (as per Census, 2011), which will further go high in the coming years. As such, the state has to produce a sufficient quantum of oilseeds for meeting the escalating demand vis-a-vis bridging the gap between requirement and production of oilseeds, for which the following strategies are to be taken up:

10.5.1 Area Expansion

1. Substitution of traditional crops with soybean and groundnut during kharif in the western red and lateritic zone;
2. Sunflower cultivation in rice-fallows in the coastal saline areas;
3. Cultivation of toria, linseed and niger after the harvest of autumn and winter rice;
4. Area expansion under sesame in all the three seasons in both traditional and non-traditional areas;
5. Expansion of area under rabi/summer groundnut in rice-fallows and on residual moisture in flood plains of Assam, West Bengal, Orissa and Chhatishgarh;
6. Intercropping/mixed cropping soybean with maize, sunflower with groundnut/pulses and linseed with wheat/pulses;
7. Cultivation of mustard in rice-fallows in north-eastern states;
8. More cultivation of sesame as a summer crop after rice in eastern states;
9. Cultivation of soybean and sunflower in upland areas of Orissa, Tamil Nadu, Bihar and Andhra Pradesh as these are more remunerative than upland rice.

Table 10.5 Area, production and productivity of total oilseeds in West Bengal. (Sources: Anonymous 2010, 2011)

Year	Area ('000 ha)	Production ('000 t)	Productivity (Kg ha^{-1})
2000–2001	598.55	570.67	953
2001–2002	604.15	493.04	816
2002–2003	568.39	475.83	837
2003–2004	685.05	651.74	951
2004–2005	673.15	556.82	827
2005–2006	643.47	623.30	969
2006–2007	703.41	645.38	917
2007–2008	707.43	705.71	998
2008–2009	703.68	582.62	828
2009–2010	681.99	726.19	1066

10.5.2 Improving Productivity

1. Replacement of varieties with newer ones;
2. Use of quality seeds with optimum seed rate;
3. Timely adoption of improved cultural practices;
4. Balanced fertilization including micronutrients and gypsum application as a source of sulphur;
5. Transfer of technology to the farmers;
6. Adoption of resource conserving technologies;
7. Exploitation of other non-traditional sources of oil (e.g. oilpalm).

The state of West Bengal is endowed with favourable agro-climatic conditions for the production of a variety of crops throughout the year. Though limited scope for crop diversification prevails during kharif, there is great scope of diversification both under irrigated and rainfed conditions during rabi and summer seasons in West Bengal (Samui 2005). Owing to huge deficit in the production of oilseeds in the state, there is a dire need to emphasize increased production of oilseeds through diversification in traditional cropping systems, introduction in non-traditional areas and non-conventional seasons utilizing residual moisture, and improving productivity with right application of available technology (Anonymous 2002). Similar strategies need to be adopted for other eastern states of the country.

10.6 Suitability of Oilseeds in Different Cropping Systems

The rainfall receipts in eastern India are usually high and some amount of rainfall is received during pre-kharif (March–May). A fairly good amount of moisture remains stored in the soil after the harvest of kharif crops. Further, the water table is sufficiently high, between 1–2 m below the ground level in the medium lands in many areas. All these facilitate in growing more than two crops in these predominantly rice-based mono-cropped areas, even without irrigation facilities. Oilseed crops have the potential for increasing cropping intensity and profitability in a wide range of cropping systems. These crops by nature are hardy, mostly grown under rainfed conditions, and impart stability of production system under harsh conditions (Reddy and Suresh 2009). As most of them have deep roots, they can tap very well the untapped resource of soil moisture (Chatterjee 1985). Moreover, these crops by virtue of their low irrigation requirement and better remunerative price are ideally suited to replace low-yielding other crops and become popular even in non-traditional areas (Palaniappan and Jeyabal 2002). Therefore, adapting cropping systems with oilseed crops in non-traditional areas and seasons can enhance the oilseed production substantially. But, these cropping systems require more balanced supply of inputs and effective crop management over mono-cropping because of the higher nutrient uptake by the crops and competition for light, water and nutrients (Palaniappan and Jeyabal 2002). Possibilities exist for expanding oilseeds cultivation both in irrigated and rainfed areas with the introduction of suitable oilseed crops, which can be intercropped, mixed cropped, relay cropped or grown as catch crop in multiple sequences (Bhowmick et al. 2006).

10.6.1 Sequential and Relay (Paira/Utera) Cropping

As the prices of oilseeds have gone high, the farmers can grow oilseeds as a cash crop. Time has come to provide them with cropping sequences, which can be stated as "oilseed-based cropping sequences" instead of "cereal-based cropping sequences". These cropping sequences have profound influence on the productivity of succeeding crops (Chatterjee 1985).

Rapeseed being mostly taken as a catch crop enjoys the facility of being grown in solid stand (Sengupta and Das 2003). If grown after aman rice, long-duration varieties do not suit due to short and mild winter in West Bengal during rabi, and aman rice varieties also should be of shorter duration. Paira cropping of rapeseed-mustard in between two rice crops in rice–rice system is now gaining popularity (Samui 2005). Paira cropping of yellow sarson in the standing aman paddy has been found economic when the seeds of paira crop were sown broadcast before rice harvest. Broadcasting seeds of yellow sarson after land preparation in rice-fallow was found uneconomic due to severe infestation of aphid and diamond back moth and the prevalence of high temperature at later stage of crop growth due to delayed sowing in West Bengal (Samui 2005). Similarly, linseed can be grown as paira crop (Chatterjee 1985). Growing of mustard and linseed after rice was found remunerative (Yadav and Shukla 2002).The principle behind relay/utera/paira cropping is to take advantage of the residual moisture to raise two crops in a year by adopting an overlapped cropping system (Kempanna 1985). The areas in which this system is widely prevalent are West Bengal, parts of Orissa, Uttar Pradesh and Madhya Pradesh, Bihar and Chhatisgarh (Bhowmick et al. 2006).

Among the oilseeds, groundnut and sesame can be cultivated during summer season in aman fallow areas, where even restricted irrigation facilities are available. These two crops gave much higher yield with less disease incidence when grown in pre-kharif than during kharif. Groundnut and sunflower being fairly resistant to soil salinity have already become popular second crops in aman fallows of coastal tracts in West Bengal where cultivation of any second crop was earlier not possible (Chakraborty et al. 1973). Cultivation of rabi groundnut in river beds of West Bengal on residual moisture was found remunerative and cost-effective. The scope of groundnut cultivation is also replacing non-remunerative upland rice during kharif in upland and medium lands under red and lateritic zones of West Bengal (Samui 2005).

In West Bengal, pre-kharif sesame is mostly grown after harvest of potato, aman rice or jute. Potato being grown generally in light and fertile soils requires frequent irrigation, allowing a considerable amount of Nitrogen to leach down the root zone. The subsequent sesame crop having comparatively deeper root system is expected to draw nutrients from a deeper zone where potato roots cannot reach. When grown after potato, no fertilizer is therefore applied to sesame as the residual fertility seems to be enough for good crop growth (Chatterjee et al. 1992; Sengupta and Das 2003). However, in such rotation, sesame is found to be affected by Macrophomina disease due to high soil fertility status, encouraging profuse vegetative growth. Jute–sesame rotation is comparatively safer (Sengupta and Das 2003).

The usual crops preceding safflower are greengram, blackgram, maize, millets, sesame, upland paddy, groundnut, soybean and cowpea (Sengupta and Das 2003). Inclusion of leguminous oilseed in the cropping system has several advantages. Root nodules of legume release N during decomposition for the use of succeeding crops (Kailasam 1988). Prasad (1996) also reported that nitrogen fixed by the legumes not only meets their own N requirements but also a sizeable quantity (30–90 kg ha^{-1}) is left for the succeeding crops. Mustard being highly responsive to N-application yields more after leguminous crops, more particularly after cowpea taken as fodder. Studies also indicate that sunflower is a less exhaustive crop than maize, but a time gap of at least 2 weeks should be allowed between the two crops (Palaniappan and Jeyabal 2002).

In the new alluvial zone of West Bengal, cropping sequence of rice–wheat–groundnut was as highly productive as rice–potato–jute (Katyal and Gangwar 2001). Similarly, rice–potato–groundnut and rice–vegetables–sesame were found productive, profitable and efficient in irrigated areas (Samui et al. 2004; Samui 2005). Sharma and Rajput (1990) found rice–mustard–sesame and rice–mustard–greengram as more water-efficient systems at Memari (Burdwan) in West Bengal. In Bihar, rice–potato–sunflower system recorded higher rice-equivalent yield, net returns, benefit–cost ratio, land-use efficiency and production efficiency than traditional rice–wheat-green manuring (Reddy and Suresh 2009).

Under coastal rice-fallows, mainly rabi oilseeds like sunflower, safflower, linseed and castor are grown owing to mild temperature during winter (Ali and Mishra 2002). Again, instead of existing rice–rice system in coastal areas, rice–potato–sesame, rice–rice–soybean and rice–groundnut have been identified to be more suitable with high productivity, profitability and stability for the coastal areas of Orissa, Tamil Nadu and Maharashtra, respectively (Gangwar et al. 2003). Under the situation of limited irrigation availability in central India, soybean–wheat and soybean–chickpea were found most efficient and stable (Gangwar et al. 2004).

However, some cropping sequences, where oilseed crops can easily and profitably be nurtured, are (1) irrigated upland situation: (i) direct seeded rice (aus)/ jute–mustard–reengram/ sesame, (ii) groundnut/sesame–mustard–greengram, (iii) direct seeded/transplanted rice (aus)–rapeseed–groundnut; (2) irrigated medium and low land situations: (i) rice (direct seeded/ transplanted)/jute–rapeseed/mustard–boro rice, (ii) direct seeded rice–sesame (short duration)–wheat, (iii) direct seeded rice–wheat/potato/vegetables–sesame/groundnut; (3) other sequences suitable for irrigated condition: (i) groundnut-sesame/greengram, (ii) cotton–sunflower, (iii) rice–sunflower–greengram, (iv) sunflower–chickpea/soybean, (v) cowpea–sunflower–maize, (vi) sunflower–rice–sesame/blackgram/greengram, (vii) groundnut–wheat/maize, (viii) rice–niger (pre-rabi)/sunflower, (ix) rice–potato–sunflower, (x) soybean–sunflower; (4) rainfed upland and medium land situations: (i) jute–direct seeded rice–rapeseed/mustard/safflower, (ii) direct seeded rice (early)–niger/ safflower, (iii) jute–direct seeded/transplanted rice–linseed/mustard/safflower/sunflower; (5) rainfed lowland situation: (i) rice (late variety)–sesame, (ii) groundnut–sorghum/bajra/maize/sesame, (iii) millets/cowpea/kharif pulse–sunflower, (iv) sunflower–safflower.

10.6.2 Intercropping and Mixed Cropping

The underlying principle, regardless of crops and varieties, is to distribute/dilute the risk arising out of aberrant weather besides enhancing farm income. The basic objectives are crop intensification in both time and area, and raising productivity per unit of area and input by increasing the pressure of plant population. It also entails better utilization of soil moisture, nutrients and solar radiation than sole cropping of the base crop. The important criterion is not to sacrifice the main crop yield, but at the same time to have some additional yield. Some prominent examples are as follows:

Rapeseed-mustard Mustard is often mixed cropped or intercropped with sugarcane, wheat, potato and gram. In irrigated areas, mustard as companion crop of autumn planted sugarcane (Chatterjee 1985; Chatterjee and Pal 1988; Yadava et al. 2001; Palaniappan and Jeyabal 2002; Sengupta and Das 2003) in the sub-tropical region has been found to be more remunerative than pure sugarcane. Sugarcane is planted in 90 cm apart rows and two rows (30 cm apart) of mustard can easily be sown in between (Chatterjee 1985).

When shallow-rooted wheat was intercropped with deep-rooted mustard and chickpea, the consumptive use of water and combined intercrop yield increased because of more efficient use of soil moisture from the deeper zones (Mandal et al. 1986). One row of mustard after every ninth row of wheat proved to be the appropriate method (Chatterjee and Pal 1988; Yadava et al. 2001) under irrigated condition. The mustard variety Sanjukta-Asech (RW-4C-6–3/II) is suitable for mixed cropping with wheat in West Bengal.

Because of dissimilar growth, duration and canopy characteristics, intercropping of mustard (Varuna) with potato (Kufri Chandramukhi) in 1:3 proportion by border-ridger technique was found superior over pure crops (Chatterjee 1985; Yadava et al. 2001). Here, mustard should receive the first irrigation after earthing up of potato and subsequent irrigations to potato should be given after closing furrows from both the sides of the ridge on which mustard is planted. Intercropping of mustard (Varuna) and chickpea in a 1:2–3 row proportion has been found to be beneficial both under irrigated and rainfed conditions. Intercropping of lentil and toria in a 1:1 proportion was found suitable for rainfed areas of Uttar Pradesh (Yadava et al. 2001).

Sunflower It can profitably be intercropped with fingermillet, pigeonpea, groundnut, soybean, greengram, blackgram, horsegram and cowpea. Availability of high-yielding cultivars of sunflower with varied duration, photoperiod insensitivity and wide adaptability to varied soil and agro-climatic situations has made it highly suitable for intercropping. The drought tolerant nature of sunflower has made it better suited for the intercropping systems to exploit the resources over time (Palaniappan and Jeyabal 2002). Taking sunflower as an intercrop in groundnut in a 2:6 row proportion could bring higher returns and more income stability (Chatterjee and Bhattacharyya 1986; Chatterjee et al. 1989; Sindagi and Virupakshappa 1993). More sunflower yields with less disease incidence could be obtained by alternating sunflower with soybean (Palaniappan and Jeyabal 2002).

Soybean Soybean, when introduced as intercrop, exerts less competition to companion crop; it does not impose any allelopathic effects on companion crop and helps in fertilizer economy (Reddy and Suresh 2009). It offers much scope for intercropping in sugarcane by which the fertilizer N requirement could be cut down to some extent (Kailasam 1988). Intercropping of soybean sown 60 cm apart with two rows of direct-seeded rice (20 cm apart) in between is feasible and more remunerative. The mixture of soybean sown 60 cm apart with maize rows in between soybean out yielded pure crops in both dry and wet seasons. Similarly, the mixture of soybean in rows 45 cm apart and sorghum in between soybean rows out yielded pure crops in dry season (Chatterjee and Roquib 1977). Highest monetary returns were obtained when soybean was intercropped with maize in alternate rows where each of both the crops was spaced 60 cm apart (Sen 1982).

Groundnut Being a long-duration crop, groundnut can be grown along with sunflower so that early season rains benefit sunflower and late rains benefit groundnut, thereby giving some assurance to the dry-land farmers. Advancing sowing of groundnut by 15–30 days prior to sunflower reduces the competition due to the shading of sunflower. Research data recorded a net return of ₹ 12,615 ha^{-1} under sunflower + groundnut intercropping system in West Bengal (Reddy and Suresh 2009).

Yield advantages have been reported up to 81 % when one row of groundnut was alternated with one row of sunflower (Morden) and sown in June in the light sandy loam upland soils of IGPs. Similarly, when one row of sesame (B 67) established in June was alternated with two rows of groundnut (AK 12–24), the production per unit area increased appreciably in West Bengal (Chatterjee 1985).

In an intercropping of one row of pearlmillet and three rows of groundnut with inter-row spacing of each species in mono- as well as intercrop, Marshall and Willey (1983) recorded 28 % more total dry matter and lesser risk than growing the individual crops (Chatterjee and Bhattacharyya 1986; Chatterjee and Mandal 1992). In Sundarban area of West Bengal (coastal saline tract), yield advantage was found quite high when three rows of groundnut were alternated with three rows of chillies (Reddy and Suresh 2009).

Niger Intercropping of niger either with groundnut or sunflower has been found remunerative when sown at the later part of the rainy season in West Bengal. Intercropping of niger and sunflower in a 1: 1 or 2: 1 row proportion has been found to improve the seed filling of sunflower heads due to increased bee population (Chatterjee 1985). Niger can also be mixed cropped with bajra, ragi, castor, groundnut or pulses under rainfed situations during kharif and rabi (Anonymous 2001).

Linseed It can be grown as mixed crop/intercrop during rabi season with gram, wheat or lentil in a 1:3 or 3:1 proportion (Chatterjee 1985; Anonymous 1995). Inter- or mixed cropping of linseed with gram appeared to be more profitable than their sole cropping and this could appreciably reduce the wilt disease incidence and borer damage in chickpea (Chatterjee 1985; Chatterjee et al. 1992; Anonymous 1995). The fertilizer needs of chickpea and lentil are also minimized when they are intercropped with linseed. As far as possible, mixed cropping should be discouraged and intercropping should be popularized (Anonymous 1995). Moreover, intercropping of pulses and oilseeds with a recommended planting pattern after rice enhanced the net returns and improved soil health. The highest net return was recorded in rice-chickpea + linseed, closely followed by rice-linseed + mustard (Tripathi and Rathi 2003).

Sesame In West Bengal, a single row or twin row (20 cm apart) of sesame can easily be accommodated in the inter-row space of spring planted sugarcane. This practice might help in the harvesting of about 5 q additional seed yield of sesame ha^{-1} in addition to almost normal yield of sugarcane from the same piece of land. In the twin row system, the first row of sesame should be planted 35 cm away from the adjacent row of sugarcane. Similarly, a twin row of sesame (20 cm apart) can be successfully accommodated creating a space by skipping a row of spring/summer groundnut after every fourth row and such intercropping might give around 1.5 q additional oil yield ha^{-1} (Chatterjee et al. 1992).

Safflower Spices, legumes, oilseeds and cereals are the crops chosen for the intercropping of safflower. The ratio between the main crop and safflower varies from 2:1 to 3:1. In Madhya Pradesh only, the adopted ratio for intercropping safflower with either blackgram or linseed is 2:1 or 6:2 (Maity et al. 1988).

Castor Castor is usually raised either as sole or inter/mixed crop with kharif cereals/millets (sorghum, fingermillet, pearlmillet, maize)/grain legumes (pigeonpea, groundnut, mungbean, urdbean and cowpea) and sometimes with horticultural

Table 10.6 Remunerative intercropping systems with their row proportions. (Sources: Reddy and Pati 1998; Palaniappan and Jeyabal 2002)

Intercropping system	Main crop to intercrop ratio
Sugarcane + rapeseed-mustard (early)	1:2
Wheat + mustard	9:1
Potato + mustard/yellow sarson	3:1
Gram + mustard	3:1 or 2:1
Lentil/sugarcane/gobhi sarson + toria	1:1
Pigeonpea + sunflower	1:1 or 1:2
Soybean + sunflower	1:1
Groundnut + sunflower	6:2
Blackgram/greengram + sunflower	3:1 or 4:2
Direct seeded rice + soybean	2:1
Groundnut + pearlmillet	3:1 or 6:2
Groundnut + sunflower	2:1 or 3:1 or 6:2
Groundnut + pigeonpea/blackgram	4:1
Groundnut + sesame	2:1
Groundnut + castor	5:1 or 6:2 or 3:1
Niger + blackgram	2:2
Niger + sunflower	1:1 or 2:1
Niger + ragi	6:3
Niger + mungbean	1:1
Sesame + greengram/pearlmillet/groundnut	1:1
Linseed + lentil	1:3
Linseed + gram	2:1 or 3:1
Linseed + safflower	2:1
Linseed + wheat	1:3 or 3:1
Wheat + safflower	2:1 or 3:1
Coriander/chickpea + safflower	3:1
Castor + pigeonpea	1:1
Castor + greengram/blackgram	1:2
Castor + groundnut	1:3

crops like chillies, turmeric, ginger, Dolichos and cucumber (Raghavaiah et al. 2005; Reddy and Suresh 2009).

The above findings (Table 10.6) clearly indicate that with the remunerative intercropping systems, not only can more production be achieved, but production can also be more stable over the years, particularly where land and irrigation resources are limited (Bhowmick et al. 2006). Thus, oilseeds have an edge over other crops in price, wider adoptability and relative optimal production under environmental stress condition.

10.7 Conclusion

The present review clearly indicates the advantages of crop diversification with oilseeds in marginal ecosystems to make the cropping enterprise a profitable venture avoiding risks. Hence, crop diversification with oilseeds can be used as a tool for maximizing productivity and profitability leading to resource conservation, for which concerted efforts are required to step up oilseed production through both area expansion and productivity increase for making India self-sufficient in vegetable oils. Oilseeds, having the attributes to produce deep roots in order to tap the moisture from lower depths of soil, offer a great promise for their successful introduction in newer niches or non-traditional and unutilized areas. Rice-fallows, especially in eastern India, can easily be converted to double-cropped areas by introducing suitable oilseed crops like rapeseed-mustard, linseed, niger, sesame, groundnut, sunflower and safflower either as a sole or as an intercrop. Being fairly tolerant to soil salinity, cultivation of groundnut, sunflower and safflower can be expanded in the coastal areas, where other crops or other oilseeds do not grow so well. Increasing irrigated area is also worth considering diverting some cereal areas or low-yielding/less profitable crop areas in favour of oilseeds, the yields of which can be doubled or trebled through proper management under irrigated condition. Even in rainfed areas, the production of oilseeds can be sufficiently enhanced through better management practices. In the IGPs, where cereal-based cropping system is predominant, the potentiality of leguminous oilseeds like soybean and groundnut may be exploited. Exploitation of perennial sources of oilseeds including minor or non-traditional oil-bearing species, and development of strong and vibrant programmes for quality seed production and supply are some other avenues in this regard.

More research is, however, necessary to evolve location-specific, pest-resistant and short-duration varieties with high harvest indices to fit them well in different cropping systems and to introduce in non-traditional areas and seasons. Besides, there is a need to give greater impetus on refinement and/or upgradation of production technology. The development of cost-effective or high yield-low cost technology can change the scenario for oilseeds production in the country.

References

Ali M (2004) Role of pulses in crop diversification. In: Ali M, Singh BB, Kumar S, Dhar V (eds) Pulses in new perspective. ISPRD, IIPR (ICAR), Kanpur, pp 245–259

Ali M, Kumar S (2002) Diversification in cropping systems through pulses. Proceedings of National Symposium. In: Ali M, Chaturvedi SK, Gurha SN (eds) Pulses for sustainable agriculture and nutritional security. ISPRD, IIPR (ICAR), Kanpur, pp 13–22

Ali M, Mishra JP (2002) Exploiting rice fallows through oilseeds and pulses. In: Rai M, Singh H, Hegde DM (eds) Oilseeds and oils-research and development needs. Indian Society of Oilseeds Research, Directorate of Oilseeds Research (ICAR), Hyderabad, pp. 84–93

Anonymous (1995) Linseed: package of practices for increasing production (1st revised edition). Directorate of Oilseeds Research (ICAR), Hyderabad, pp 1–17

Anonymous (2001) Technology for increasing production of niger. In Singh NB, Duhoon SS (eds) Tech. Bulletin No. 2. AICRP on Sesame and Niger (ICAR), Jabalpur, pp. 1–21

Anonymous (2002) Annual plan (2002–03) on agriculture containing achievements of ninth five year plan (1997–98–2001–02) and targets of tenth five year plan (2002–03–2006–07). Department of Agriculture, Government of West Bengal, West Bengal

Anonymous (2005) State budget 2005-alternative approach to development, West Bengal XLVII (4) (April issue), pp 9–27

Anonymous (2010) Estimates of area and production of principal crops in West Bengal (2008–2009). Evaluation Wing, Directorate of Agriculture, Government of West Bengal, 17 S. P. Mukherjee Road, Kolkata, West Bengal, p 25

Anonymous (2011) Bringing green revolution in green Bengal. Directorate of Agriculture, Government of West Bengal

Bhowmick MK, Dutta A, Saha PK, Gupta S, Man GC, Ghosh RK (2006) Crop diversification through oilseeds in India with particular reference to West Bengal. SATSA Mukhapatra—Ann Tech Iss 10:37–61

Chakraborty LN, Sen SN, Mandal SK, Sen Gupta K, Mukherjee D (1973) Possibility of utilizing rice fallow in West Bengal. In: Das Gupta DK (ed) Proceedings of Sem on possibility of growing a second crop after rice in West Bengal, Calcutta, pp 71–76

Chatterjee BN (1985) Oilseed production in eastern India, with particular reference to West Bengal. In: Srivastava HC, Bhaskaran S, Vatsya B, Menon KKG (eds) Oilseed production: constraints and opportunities. Oxford & 1BH Pub. Co., New Delhi, pp 347–358

Chatterjee BN, Bhattacharyya KK (1986) Principles and practices of grain legume production. Oxford & IBH Pub. Co., New Delhi

Chatterjee BN, Maiti S, Mandal BK (1989) Cropping systems-theory and practice (2nd edn). Oxford & IBH Pub. Co., New Delhi

Chatterjee BN, Mandal BK (1992) Present trends in research on intercropping. Indian J Agri Sci 62(8):507–518

Chatterjee BN, Roquib MA (1977) Soybean research (1967–1975). Bulletin No. 3, B. C. K. V., West Bengal, pp 1–50

Chatterjee SD, Pal SR (1988) Rapeseed-mustard: technologies for better crops. Govt. of West Bengal, Directorate of Agriculture, Pulses and Oilseeds Research Station, Berhampore

Chatterjee SD, Bhattacharya B, Chatterjee AK, Jha AR, Shrivastava MP, Laha SK, Roy S (1992) Sesame and linseed: technologies for better crops. Government of West Bengal, Directorate of Agriculture, Pulses and Oilseeds Research Station, Berhampore

Gangwar B, Prasad K (2005) Cropping system management for mitigation of second-generation problems in agriculture. Indian J Agri Sci 75(2):65–78

Gautam RC, Sharma AR (2004) Diversification in cereal-based cropping systems for sustained productivity and food seurity. Indian Farming (Oct. issue), pp 3–8

Gangwar B, Katyal V, Anand KV (2003) Different cropping systems for coastal areas. J Indian Soc Coastal Agri Res 21:21–30

Gangwar B, Katyal V, Anand KV (2004) Stability and efficiency of cropping systems in Chhattisgarh and Madhya Pradesh. Indian J Agri Sci 74:521–528

Gill MS, Ahlawat IPS (2006) Crop diversification—its role towards sustainability and profitability. Indian J Fert 2(9):125–138 and 150

Hegde DM (2000a) Strategic issues and technological options for doubling oilseeds production. Souvenir. Natl. Sem. On "Oilseeds and oils-R & D needs in the millennium", Feb. 2–4, 2000, Indian Society of Oilseeds Research, Hyderabad, pp 1–10

Hegde DM (2000b) Oilseeds-technology for high yields. The Hindu Survey of Indian Agriculture, pp 65–71

Hegde DM (2002) Oilseeds-Measures to turn self-reliant. The Hindu Survey of Indian Agriculture, pp 71–76

Hegde DM (2004) Oilseeds-Becoming self-reliant. The Hindu Survey of Indian Agriculture, pp 45–47

Hegde DM (2005a) Oilseed scenario in India-Past, present and future with special reference to rapseed-mustard. In Proc. Winter School on "Advances in rapeseed-mustard research technology for sustainable production of oilseeds" (Dec. 15, 2004–Jan. 04, 2005). NRCRM (ICAR), Bharatpur, Rajasthan, pp 1–15

Hegde DM (2005b) Oilseeds-Striving for self-sufficiency. The Hindu survey of Indian Agriculture, pp 58–62

Joshi PK (2005) Crop diversification in India: nature, pattern and drivers. www.adb.org/Documents/Reports/Consultant/TAR-IND-4066/…/joshi.pdf. Accessed 28 Jan 2012

Kailasam C (1988) Intercropping of oilseeds in sugarcane. Natl. Sem. "Strategies for making India self-reliant in vegetable oils", Sept. 5–9, 1988. Indian Society of Oilseeds Research, Hyderabad, pp 325–332

Kalra GS (1990) Crop diversification for optimizing the use of agricultural input and improving productivity in north west India. Nat Bank News Review, Bombay 6(3):15–21

Katyal V, Gangwar B (2001) Productivity, stability and profitability of rice-based cropping sequences in West Bengal and Orissa. Indian J Agron 46:387–394

Kempanna C (1985) Prospects of horizontal growth of oilseeds production in the eighties and nineties. In: Srivastava HC, Bhaskaran S, Vatsya B, Menon KKG (eds) Oilseeds production: constraints and opportunities. Oxford & IBH Pub. Co., New Delhi, pp 73–87

Mahapatra IC (2002) Multidisciplinary problem solving demand driven location and ecosystem specific participatory rice research and development in eastern India. Presidential Address at the 40thG.B. Meeting of the ARRW, Dec. 12, 2001, CRRI, Cuttack, Orissa. Oryza 39(1–4):i–xiii

Maity S, Hegde MR, Chattopadhyay SB (1988) Hand book of annual oilseed crops. Oxford & IBH Pub. Co., New Delhi

Mandal BK, Dasgupta S, Ray PK (1986) Yield of wheat, mustard and chickpea grown as sole crop and intercrop with 4 moisture regimes. Indian J Agri Sci 56(8):577–583

Marshall B, Willey RW (1983) Radiation interception and growth in an intercrop of pearlmillet/groundnut. Field crops Res 7:141–160

Menon KKG (1985) Increasing vegetable oil production in India-we can, we must. In Srivastava HC, Bhaskaran S, Vatsya B, Menon KKG (eds) Oilseeds production-constraints and opportunities. Oxford & IBH Pub. Co., New Delhi, pp 17–30

Mishra B (2005) Rice-more crop per drop. The Hindu Survey of Indian Agriculture, pp 41–46

Palaniappan SP, Jeyabal A (2002) Crop diversification with oilseeds in non-traditional areas and seasons. In Rai M, Singh H, Hegde DM (eds) Oilseeds and oils-research and development needs. Indian Society of Oilseeds Research, Directorate of Oilseeds Research (ICAR), Hyderabad, pp. 94–100

Prasad R (1996) Cropping system and sustainability of agriculture. Indian Farm 46:39–44

Raghavaiah CV, Lavanya CL, Raoof MA, Lakshminaryana M (2005) Castor: package of practices for increasing production. 4th revised edition (Hegde, D. M., ed.), Directorate of Oilseeds Research (ICAR), Hyderabad, pp 1–20

Rai M (2002) Oilseeds on a balance. In Rai M, Singh H, Hegde DM (eds) Oilseeds and oils-research and development needs. Indian Society of Oilseeds Research, Directorate of Oilseeds Research (ICAR), Hyderabad, pp 1–36

Rao MV (2008) Oilseeds—several constraints in increasing productivity. The Hindu Survey of Indian Agriculture, pp 39–41

Reddy PS, Pati D (1998) Indian oilseeds: present status and future needs. Indian J Agri Sci 68(8, Spl Issue):435–459

Reddy BN, Suresh G (2009) Crop diversification with oilseed crops for maximizing productivity, profitability and resource conservation. Indian J Agron 54(2):206–214

Samui RC (2005) Role of cropping system in crop diversification and sustainable crop production. SATSA Mukhapatra-Annual Technical Issue 9:12–15

Samui RC, Kundu AL, Majumdar D, Sahu PK (2004) Diversification of rice (*Oryza sativa*)-based cropping system in new alluvial zone of West Bengal. Indian J Agron 49:71–73

Sarkar RK (2005) Diversification of crops and cropping system for nutritional security. SATSA Mukhapatra-Ann Tech Iss 9:1–4

Saxena S, Agarwal RC, Sapra RL, Singh AK, Gautam PL (2002) Biodiversity and information management system. In: Rai M, Singh H, Hegde DM (eds) Oilseeds and oils-research and development needs. Indian Society of Oilseeds Research, Directorate of Oilseeds Research (ICAR), Hyderabad, pp 55–67

Sen SN (1982) A review of research work on soybean done under the Directorate of Agriculture, West Bengal. In Pulses and Oilseeds in West Bengal. Directorate of Agriculture. Govt of West Bengal, pp 57–71

Sengupta K, Das PK (2003) Cultivated annual oilseed crops of India (1st edn). Naya Udyog, Kolkata, 295 p

Sharma BR, Rajput AL (1990) Sustained rice productivity in eastern region of India through efficient water management technologies. Indian J Agron 35:85–90

Siddiq EA (1999) Rice-not a distant dream. The Hindu survey of Indian Agriculture, pp 39–46

Sindagi SS, Virupakshappa K (1993) Sunflower. ICAR, Krishi Bhavan, New Delhi, pp 1–36

Tripathi AK, Rathi KS (2003) Productivity, profitability and sustainability of Indian mustard (*Brassica juncea*) raised after different rainy-season crops. Indian J Agron 48:251–253

Yadava JS, Singh AK, Chauhan JS (2001) Strategic issues for doubling the productivity of rapeseed-mustard in next decade in India. Ann Biol 17(1):1–20

Yadav RL, Shukla ND (2002) Diversification in cropping systems for sustained production of oilseeds. In Rai M, Singh H, Hegde DM (eds) Oilseeds and oils-research and development needs. Indian Society of Oilseeds Research, Directorate of Oilseeds Research (ICAR), Hyderabad, pp 101–111

Zaman A (2012) Integrated water resource management for sustainable agricultural production in eastern India. Abstracts: lead Papers and Research Articles. Natl. Sem. On "Safe Food For All", Feb. 21–23, 2012, Department of ASEPAN, Palli Siksha Bhavana (Institute of Agriculture), Visva-Bharati, Sriniketan, West Bengal, India. pp I–20–I–21

Chapter 11
Crop Diversification Through Pulses in the Northeastern Plain Zone of India

P. K. Biswas and M. K. Bhowmick

11.1 Introduction

Being comparatively of short duration and having the ability to thrive better than other crops under harsh climate and fragile ecosystems on one hand and to achieve household food and nutritional security on the other, pulses offer great promise towards diversification of cropping systems. They can be intercropped, relay cropped or grown as catch crop in multiple sequences. Under rainfed-upland ecosystem, monocropping of rice is generally done, as the soils of the ecosystem are poor in fertility and have low moisture retention capacity. The rainfed lowlands, where nutrient status and moisture availability are enough, may support short-duration crops if managed properly (Ali and Kumar 2002). Short-duration pulses appear to be potential crops for double cropping under such ecosystem.

Increasing demand for pulses to meet the requirement of burgeoning population has brought the non-traditional and utilized areas in sharp focus. Moreover, inclusion of food legumes in rice production system not only brings additional area under the crops but also improves the physical, chemical and biological properties of soils. Thus, sustainability of the crop production system vis-à-vis urgency of enhancing pulses production call for immediate corrective steps in the form of crop diversification (Bhowmick et al. 2005). An important step in this direction is to grow suitable *rabi* pulses with improved package of practices.

P. K. Biswas (✉)
Institute of Agriculture (Palli Siksha Bhavana), Visva-Bharati, Sriniketan, Birbhum, West Bengal 731236, India
e-mail: ppabitra07@rediffmail.com

M. K. Bhowmick
Rice Research Station (Government of West Bengal), Chinsurah (R.S.), Hooghly, West Bengal 712102, India
e-mail: bhowmick_malay@rediffmail.com

M. Ghosh et al. (eds.), *Diversification of Agriculture in Eastern India*,
India Studies in Business and Economics, DOI 10.1007/978-81-322-1997-2_11

11.2 Crop Diversification: The Concept

Diversified cropping refers to a cropping plan in which no single crop contributes 50% or more towards the total crop production or monetary income (comparable equivalents) annually (Anon 1996). Crop diversification is intended to give a wider choice in the production of a variety of crops in a given area so as to expand production-related activities on various crops and also to lessen risk. Crop diversification in India is generally viewed as a shift from traditionally grown less remunerative crops to more remunerative crops. The crop diversification also takes place due to governmental policies and thrust on some crops over a given time. Market infrastructure development and certain other price-related supports also induce crop shift. Higher profitability and the resilience/stability in production also induce crop diversification. Crop diversification and the growing of a large number of crops are practiced in rainfed lands to reduce the risk factor of crop failures due to drought or less rain. Crop substitution and shift are also taking place in the areas with distinct soil problems. For example, growing of rice in high water table areas replacing oilseeds, pulses and cotton, promotion of soybean in place of sorghum in vertisols, etc. (Hazra 2012).

11.3 Nature of Rice-Fallows

In India, approximately 15 million ha (m ha) area is under rice-fallow ecosystem (Hegde 1999) which can be utilized for successful cultivation of pulses with selection of appropriate varieties of rice and succeeding pulses besides improved management practices. This calls for critical analysis of production constraints and appropriate remedial measures, particularly in the northeastern plain zone (NEPZ), comprising eastern Uttar Pradesh (UP), Assam, Bihar, Orissa and West Bengal (WB). Rice-fallows in this zone can be considered as possible new niches for pulses area expansion (Table 11.1). The common features of rice-fallow ecosystem of NEPZ are as follows:

1. Higher soil moisture at the time of harvesting of rice in low lands and moisture stress in upland conditions is generally observed. The succeeding crop, however, invariably suffers from moisture stress at the terminal phase of their life-cycle.
2. During winter, *rabi* pulses can be successfully cultivated. Frost and foggy weather is often encountered during winter season.
3. The soils are alluvium (new and old) with neutral to alkaline in reaction having high contents of calcium carbonate and sandy to clayey in texture.
4. The organic matter content is generally low to medium with low moisture retention capacity and of low fertility status with micronutrient deficiency in many pockets (Ali and Mishra 2002).

Table 11.1 Rice-fallows for pulses production. (Sources: Ali 1999 and Asthana 1999)

Crop	Agro-climatic zone	Specific areas	Potential areas in m ha (2020 A.D.)
Chickpea	Sub-humid to humid zone	East Uttar Pradesh, Bihar, West Bengal and coastal Orissa	1.0
Urdbean (*Rabi*)	Coastal areas of semi-arid zone	Andhra Pradesh, Karnataka and Tamil Nadu	2.0
Lentil	Sub-humid to humid semi-arid zone	East Uttar Pradesh, Bihar, Bundelkhand of Uttar Pradesh and Madhya Pradesh	1.0
Lathyrus	Sub-humid to humid zone	East Uttar Pradesh, Bihar, Orissa, West Bengal and Chhattisgarh	1.4

11.4 Suitability of Pulses for Diversification in Rice-Based Systems

There are several crop sequences (Table 11.2) in which rice and legumes (pulses) are cultivated (Pravakaran and Rangarajan 2010) either as a sequential (catch) crop or as a relay (*paira*) crop. In case of *paira* (or *utera*) cropping, the crop is sown by broadcasting in the standing crop of lowland rice before its harvest, where the residual moisture is used for the establishment of *utera* crop; for example, lathyrus, gram, lentil, fieldpea etc. (Chakraborty et al. 1973). Here, the crops may not necessarily be sown and harvested at exactly the same time but they are usually simultaneous for a significant part of their growing periods, compared with relay cropping (*paira*), where overlapping of crops is brief (Chatterjee and Mondal 1992). *Paira* cropping is usually done under the following situations (Bhowmick et al. 2005):

1. Where late *aman* paddy is harvested by the end of November or early December on lowland heavy soils, the soil is not immediately good for ploughing due to high soil moisture condition; late sowing of crop may not be good for productivity and the soil moisture may limit a good growth late in the season (Chatterjee and Bhattacharyya 1986).
2. Where the land vacated by rice has usually no irrigation resource (Chatterjee and Bhattacharyya 1986), *paira* (relay) cropping in such unirrigated areas helps to make the best use of residual moisture of rice fields (Ali et al. 1993).
3. Where continuous cropping of rice and wheat has led to several problems such as soil sickness, deficiency of some of the nutrients, lowering of water table, salinization, etc. (Ali 1999; Siddiq 1999), introduction of pulses especially at the tail end of the command areas having limited irrigation proves to be quite promising.

Usually, the farmers, due to several reasons, do not grow pulses on irrigated lands. The basic reason is that under irrigated conditions pulses are not as remunerative as the high-yielding varieties of cereals and other commercial crops like potato,

Table 11.2 Prominent cropping systems with pulses in NEPZ of India. (Source: Pravakaran and Rangarajan 2010)

States	*Kharif*	*Rabi*	Summer
Uttar Pradesh	Rice	Gram/Lentil	Blackgram
Bihar	Rice	–	Greengram
	Rice	–	Blackgram
Assam	Rice	–	Cowpea/Greengram
West Bengal	Rice	Gram/Lentil/Lathyrus/Fieldpeas	Blackgram/Greengram

sugarcane, cotton and tobacco. This is true for several pulse crops which are mostly grown in poor and non-responsive soils, and have to survive on whatever moisture is available in the soil, either as a sole crop or as a relay or sequence crop in cereal fallows. Thus, there is a very common observation that a large diversity of cropping systems exists under rainfed areas with an overriding practice of inter- or relay cropping, while in areas with assured irrigation facilities, the major cropping systems are tilted in favour of cereals mostly (Bhowmick et al. 2005). However, high-yielding varieties of lathyrus, lentil, peas and gram can be sown in standing crop of *aman* rice when it reaches the physiological maturity, and water of rice field recedes (Sarkar et al. 1991; Mukherjee 1982).

Among *rabi* pulses, lathyrus is a significant crop of the Indo-Gangetic Plains (IGPs) and is mainly grown in Madhya Pradesh, Bihar, Chhattisgarh, Maharashtra and West Bengal. Low productivity, especially under *utera* system and presence of oxalyl diaminopropionic acid (ODAP) causing a neurotoxic disorder in monogastric animals including humans are the major problems associated with this crop. Despite these bottlenecks, lathyrus is still grown on a large area and preferred by farmers due to its hardy nature, least cost of cultivation and a good quality feed for cattle. It resists extreme harsh conditions of excessive moisture at sowing and moisture stress even at later stage of crop growth, and even late sowing does not result in corresponding yield losses (Jeswani and Baldev 1997).

Owing to the hardiness, particularly for drought, lentil is also grown under *utera* condition in rice-fallows under residual soil moisture and without much additional inputs. Rice–lentil system is very popular in eastern Uttar Pradesh, Madhya Pradesh, Bihar and West Bengal. Likewise, chickpea can also be grown. Eastern Uttar Pradesh and northern Bihar show more potential for rice–chickpea system, which is more remunerative than rice–wheat under resource constraints (Ali and Kumar 2002).

Short-duration varieties of blackgram (urdbean) and greengram (mungbean) can also give reasonable yields with the effective utilization of the residual moisture, particularly in large areas, which would have been left idle during a whole season following rice harvests in southern and eastern states. Coastal rice-fallows in the states of Andhra Pradesh, Karnataka, Tamil Nadu and Orissa can be effectively utilized for production of mungbean and urdbean during *rabi*. Kumar and Ali (1996) evaluated various pulse crops after rice and found that lentil and lathyrus were more productive and remunerative than chickpea, fieldpea and frenchbean.

11.5 Need for the Diversification of Rice-Based Cropping Systems

Though extensive testing of various combinations reveal rice–wheat and rice–rice to be the most productive among grain crop combinations, profitwise they are not as attractive as rice–pulse combinations. Besides, the following problems started appearing in the major rice-based cropping systems, especially rice–wheat in the IGPs and rice–rice in traditional rice-growing peninsular and eastern India (Siddiq 1999):

1. Declining/plateauing of factor productivity largely due to depletion of organic matter content in rice–wheat system and accumulation of organic matter coupled with slow mineralization in rice–rice system.
2. Depletion of soil fertility/productivity due to overmining of native nutrient reserve.
3. Declining fertilizer use efficiency.
4. Emergence of macro- and micro-nutrient deficiency problems in intensively cultivated areas where the soils are found deficient in N, P, K, Zn and also S.
5. Ground water depletion due to overexploitation.
6. Increasing problems of salinity–alkalinity in the command areas due to excessive and indiscriminate use of irrigation water.
7. Build-up of disease and insect pest pressure due to continuous cropping and increased varietal uniformity.
8. Emergence/resurgence of pests and their resistance to pesticides, leading to increased cost on crop protection and environmental pollution because of overdependence and indiscriminate use of pesticides; and imbalanced use of major nutrients.
9. In view of these problems, the current thinking of diversification in the contemporary cropping systems has been arisen so as to make them sustainable and profitable, besides adding insurance against the outbreak of pest incidence and socio-economic advantages. Significantly, rice or wheat rotated with a pulse (legume) is economically profitable and ecologically harmonious. The options readily available for diversification in rice-based cropping systems through pulses are: (1) inclusion of short-duration varieties of pulse crops, (2) introduction in new niches (rice-fallows), (3) substitution of existing low-yielding crops in the prevailing systems and (4) growing of pulses mainly as relay (*paira*) crop following improved management practices (Bhowmick et al. 2005).

11.6 Benefits of Growing Pulses in Rice-Based Cropping Systems

Various pulses/legumes associated with rice-based cropping systems can bring about the following benefits (Pravakaran and Rangarajan 2010):

Table 11.3 Nitrogen-fixing capacity of legumes in rice field. (Source: Pravakaran and Rangarajan 2010)

Grain legumes	N fixation from atmosphere (%)	Amount (kg N ha^{-1})
Cowpea	48–56	60–78
Mungbean	45–56	61–90
Pigeonpea	71–77	111–167

Soil Fertility Enhancement Because of the addition of organic matter through the incorporation of the above- and below-ground portions of the legumes, soil fertility status including organic fraction is enhanced.

Nitrogen Addition Legumes by their potentiality to fix nitrogen from the atmosphere add nitrogen to benefit the standing crop and the succeeding crop. In general, nitrogen-fixing capacities of different pulses (legumes) vary depending upon the kind and genotype of legume, soil type and other environmental conditions under which they are grown. The amount of nitrogen fixed ranged from 0 to 450 kg N ha^{-1} (Table 11.3).

Increased N Uptake in Rice When legumes are grown in rice solely for green manure production, the succeeding rice crop takes up more nitrogen and reduces the requirement of nitrogen for rice. There are several reports that rice produced more grains after a legume than after a non-legume or after a fallow, only due to increased nitrogen uptake.

Suppressing of Insect Pests and Weeds Building up of insects pests and diseases of rice is checked by raising a legume in between two rice crops or before a rice crop in the succeeding year. Moreover, the weed growth in the soil is also arrested.

11.7 Selection of Suitable Pulse Crops and Their Varieties for Crop Diversification

Short-duration varieties of pulse crops should be selected to fit well in cropping system. Improved varieties of *rabi* pulses should replace the long-duration and low-yielding varieties (Table 11.4). Small-seeded types of *lathyrus*, known as *lakhodi*, are better adapted in rice-fallows (Jeswani and Baldev 1997). Under West Bengal situation, the varieties Nirmal (B 1) and Ratan (Bio L 212) have been found promising (Bhowmick et al. 2005). Lentil varieties Asha (B 77), Ranjan (B 256), Subrata (WBL 58), Moitree (WBL 77) and Suvendu (WBL 81) have been recommended for cultivation in the state. Likewise, chickpea varieties Mahamaya 1 (B 108), Mahamaya 2 (B 115), Anuradha (WBG 39/2) and Bidisha (WBG 29) have been released in West Bengal. Other potential pulse crops suitable for cultivation in *rabi* season are *rabi* arhar, field peas and rajmash.

Table 11.4 List of suitable varieties of *rabi* pulses in NEPZ

Crop	Varieties	Average yield (q ha^{-1})	Days to maturity	Characteristics	Area of adaptation
Lathyrus	Nirmal (B 1)	15.0	120–130	Seeds small and grey, blue flower	West Bengal
	Ratan (Bio L 212)	15.3	108–116	Low ODAP, bold seeded, blue flower, tolerant to stress	NEPZ
Lentil	Asha (B 77)	14–15	120–125	Small and ash coloured spots, susceptible to *Rhizoctonia* and wilt, white flower, semi-spreading type	West Bengal, Orissa
	Mallika (K 75)	14	135	Semi-spreading plant with dark green foliage, seeds grey mottled and bold	NEPZ
	Pant L 639	12–16	136	Small seeded and moderately resistant to shattering	NEPZ
	Ranjan (B 256)	15–16	120–125	Small seeds, light brown with faint dark spots, susceptible to wilt, white flowers, spreading type	West Bengal
	Subrata (WBL 58)	18–20	120	Semi-erect, flowers violet, bold seeded, dark mottled, moderately resistant to wilt	Whole zone
	Suvendu (WBL 81)	15–18	110–115	Semi-erect plants, bluish white flowers, small oval seeds with dark mottled seed coat and shinning lustrous red cotyledons	West Bengal
	Moitree (WBL 77)	17–18	115–120	Erect plants, bluish purple flowers, small oval seeds with dark mottled seed coat and red cotyledons	NEPZ
Chickpea	Pusa 372 (BG 372)	22.0	110–125	Suitable for late sowing, moderately resistant to wilt, blight and root rot	NEPZ
	Pusa 256	NA	NA	Suitable for late-sown situation	West Bengal
	Uday (KPG 59)	20.0	110–125	Suitable for late sowing, tolerant to root rot, wilt and stunt	NEPZ
	Mahamaya 1 (B 108)	20–22	135–140	Seeds with wrinkled seed coat, brown, tolerant to wilt	West Bengal
	Mahamaya 2 (B 115)	22–24	125–130	Seeds reddish brown with smooth seed coat, moderately resistant to wilt and rust, susceptible to pod borer	West Bengal
Pea	Dhusar (B 22)	15–17	135–140	Seeds small and mottle green in colour	West Bengal

Table 11.4 (continued)

Crop	Varieties	Average yield (q ha^{-1})	Days to maturity	Characteristics	Area of adaptation
Rajmash	Uday (PDR 14)	120–130	20–22	Variegated red seeds	NEPZ
	Amber (IIPR 96–4)	120–125	20–24	BCMV resistance	NEPZ
	Utkarsh (IPR 98–5)	130–135	24–25	Cold tolerance, attractive seed colour	NEPZ

NEPZ Northeastern Plain Zone, *NA* information not available

11.8 Conclusion

Vast areas mainly in the eastern IGPs remain monocropped under medium- and long- duration rice varieties. Non-availability of irrigation water and delay in vacating the field after rice do not normally permit double cropping. Under these situations, *paira* (relay) cropping of *rabi* pulses in rice-fallows can convert the monocropped areas and thereby increase pulse production and sustain productivity of the rice-based cropping systems.

References

Ali M (1999) Productive and efficient pulse based cropping system. In ICAR-sponsored short course on "Recent improved methods for increasing pulses production", April 5–14, 1999, IIPR, Kanpur, pp 36–49

Ali M, Kumar S (2002) Diversification in cropping systems through pulses. In: Ali M, Chaturvedi SK, Gurha SN (eds) Pulses for sustainable agriculture and nutritional security (pp 13–22). Proceedings of National Symposium, April 17–19, 2001, Delhi, ISPRD, IIPR, Kanpur

Ali M, Mishra JP (2002) Exploiting rice-fallows through oilseeds and pulses. In: Rai M, Singh H, Hegde DM (eds) Oilseeds and oils: research and development needs. ISOR, Rajendranagar, pp 84–93

Ali M, Saraf CS, Singh PP, Rewari RB, Ahlawat IPS (1993) Agronomy of lentil in India. In Erskine W, Saxena MC (eds) Lentil in South Asia. Proceedings of the seminar on "Lentil in South Asia" (pp 103–125). March 11–15, New Delhi, India, ICARDA, Aleppo, Syria

Anonymous (1996) Agronomic terminology, fourth revised edition. Indian society of Agronomy, Division of Agronomy, I.A.R.I., New Delhi

Asthana AN (1999) Strategies for increasing pulses production. In ICAR sponsored short course on "Recent improved methods for increasing pulses production", April 5–14, 1999, IIPR, Kanpur, 1–22

Bhowmick MK, Aich A, Aich SS, Shrivastava MP, Gupta S, Man GC (2005) Crop diversification through *paira* (*utera*) cropping with *rabi* pulses. SATSA Mukhapatra Annual Tech Issue 9:43–60

Chakraborty LN, Sen SN, Mondal SK, Sengupta K, Mukherjee D (1973) In: Das Gupta DK (ed) Possibility of utilizing rice-fallow in West Bengal. Proceedings of the seminar on "Possibility of growing a second crop after rice in West Bengal" (pp 71–76). Sept 15, 1973, Calcutta

Chatterjee BN, Bhattacharyya KK (1986) Principles and practices of grain legume production. Oxford and IBH Publishing Company, New Delhi, pp 310–311

Chatterjee BN, Mondal BK (1992) Present trends in research on intercropping. Indian J Agric Sci 62(8):507–518

Hazra CR (2012) Crop diversification in India. FAO Corporate Document Repository. http://www.fao.org/docrep/003/x6906e/x6906e06.htm. Accessed 26 Jan 2012

Hegde DM (1999) Cropping systems involving rapeseed-mustard, safflower and linseed in different agro-ecological situations/regions. Proceedings of the National Trg on "Production technology of oilseeds in cropping system", pp 1–11

Jeswani LM, Baldev B (1997) Advances in pulses production technology. ICAR, Krishi Anusandhan Bhavan, Pusa

Kumar R, Ali M (1996) Chickpea gives profitable returns when grown in rice-based cropping systems. Annual report, IIPR, Kanpur, pp 15–16

Mukherjee D (1982) Problems and prospects of cultivation of pulses in West Bengal. In: Mukherji DK (ed) Pulses and oilseeds in West Bengal. Directorate of Agriculture, Government of West Bengal, West Bengal, pp 7–9

Pravakaran J, Rangarajan M (2010) Symbiotic nitrogen fixation in legumes associated with rice cropping systems. In: Kannaiyan S, Kumar K, Govindarajan (ed) Biofertilizer technology, Scientific Publishers, India, pp 245–252

Sarkar RK, Shit D, Samanta N (1991) Suitable crops for *aman* rice-fallows. Indian Fmg (Aug issue):27–28

Siddiq EA (1999) Rice—not a distant dream. The Hindu Surv Indian Agric:39–46

Chapter 12
Crop Diversification in West Bengal: Nature and Constraints

Dilruba Khatun and Bidhan Chandra Roy

12.1 Introduction

Agricultural diversification has emerged as an important strategy to stabilize income and enhance employment. Specifically for a small peasant-based economy, it is not possible for the small and marginal farmers to augment income only by raising the productivity of existing crops. In this situation, crop diversification can play an important role to enhance the earnings of the small farmers by introducing high-value crops like fruits and vegetables, potato, sugarcane, etc. Further, a sustained economic growth, rising per capita income and growing urbanization have shifted the consumption patterns in favour of high-value food commodities like fruits, vegetables, dairy, poultry, meat, and fish products from staple food such as rice, wheat, and coarse cereals. The demand for and supply of these commodities have grown much faster than those of food grains (Kumar et al. 2003; Joshi et al. 2004). Thus, the farmers can gain by reorganizing their production plan in this direction. Joshi et al. (2005) noted that a positive relationship exists between growth in cultivation of horticultural commodities and the proportion of smallholders. It indicates that crop diversification is more pronounced in areas having higher concentration of smallholders. Moreover, cultivation of horticultural crops being labour-intensive, suits the smallholders well who make use of their family labour force and ensure regular flow of income.

In India, during the current decades, the process of diversification has been widespread due to the combined effects of water-seed-fertilizer technology as well as some infrastructural development such as market centres, roads, transport, etc. in the countryside (Vyas 1996; Bhalla and Singh 1997).

D. Khatun (✉)
Department of Economics, Kulti College, Kulti - 713343, West Bengal, India
e-mail: dilrubakhatun@gmail.com

B. C. Roy
Department of Agricultural Extension, Agricultural Economics and Agricultural Statistics (EES), Palli Siksha Bhavana (Institute of Agriculture), Visva-Bharati, Sriniketan, West Bengal, India
e-mail: bcroy1969@rediffmail.com

M. Ghosh et al. (eds.), *Diversification of Agriculture in Eastern India*,
India Studies in Business and Economics, DOI 10.1007/978-81-322-1997-2_12

The West Bengal economy is diversifying at a faster rate than the all India level. There was substantial crop diversification in West Bengal in the 1980s and 1990s. Both wheat and potato production increased substantially, making West Bengal the second largest producer of potato in India. The more recent trend in crop diversification is towards high-value horticultural crops. West Bengal is now a major producer of vegetables accounting for around 17 % of total vegetables in the country. However, such a crop diversification is observed only in irrigated areas like Burdwan, Hoogli, Nadia, Midnapur, etc. but not in rainfed areas like Purulia or Bankura. Many studies were carried out to examine the nature, pattern, and drivers of crop diversification in West Bengal. The results revealed that the cropping pattern in West Bengal remained skewed towards boro paddy, potato, and oilseeds (especially mustard). Analysing the cropping pattern changes in West Bengal, Ghosh (2010) found that cropping pattern in terms of allocation of acreage has been skewed towards food grains though some crops like boro rice, potato, and oilseeds have emerged as the main crop for the farmers during the last 15–20 years.

De and Chattopadhyay (2010) examined the nature of crop diversification and impact of infrastructure on crop diversification. They found out that crop diversification has been persisting since 1970 though it has taking place at a slow rate. Among the varieties of crops, the growth of summer or boro paddy, potato, and mustard cultivation has been accelerating over the past three decades whereas, the cultivation of wheat, other cereals, pulses, jute, sugarcane, etc. have been declining over the years. They also found out that minor irrigation facilities supported by electricity, storage, and marketing facilities, etc. have played important roles with varied degrees for the diversification of crops. In addition, availability of fertilizer along with expansion of irrigation and agro implements through raising yield of crops helped in many cases to diversify the selected crops.

All these studies were at state and district level. The study at household level is very rare. Further livelihood groups did not receive any attention. In reality, different livelihood groups may react differently in response to any change. Each livelihood group has some specific problems and its capabilities are different. Therefore, we get a clear picture of crop diversification if the analysis is done across livelihood groups. This chapter is an attempt to analyse crop diversification in West Bengal in this direction.

The specific objectives of this study are to: (1) examine the nature and extent of crop diversification across the livelihood groups in West Bengal, (2) identify the constraints to crop diversification in the study area, and (3) suggest some policy implications of the result. The chapter is divided into four sections including the introduction. The following two sections provide a brief description of the data and analytical tools. Section 12.4 presents the findings of the study. The concluding observations and implications are given in Sect. 12.5.

12.2 Database

The study is conducted in the state of West Bengal. To fulfil the various objectives of the study, household survey and focused group discussions (FGD) were undertaken in the study area during 2008–2009. A multistage sampling technique was used to select the sample households following specific criteria. In the first stage, two districts were selected purposively, one representing a more diversified agriculture (Burdwan) and the other less diversified one (Purulia). Next, two blocks from each of the district and two villages from each of the block were selected randomly. Accordingly, the selected villages were Barampur and Debogram under Burdwan district and Narayanpur and Goaladi under Purulia district. In all, 50 households from each of the villages were selected randomly in probability proportionate to major livelihood groups in the study villages. Thus, a total of 200 households were selected for the study. All these households then categorized into seven different livelihood groups, based on their primary source of income, namely, agricultural and allied activities; agricultural labourer, non-agricultural labourer, salaried group, casual labourer, petty business, and others.

12.3 Analytical Tools

12.3.1 Measures of Diversification

The level of diversification is measured by various types of concentration and diversification indices. Six different measures of diversification viz., Herfindahl index (HI), Simpson index (SI), ogive index (OI), entropy index (EI), modified entropy index (MEI), and composite entropy index (CEI) have been used.

Herfindahl Index (HI) HI is computed by taking sum of squares of acreage proportion of each crop in total cropped area as given by the following formula:

$$\mathrm{HI} = \sum_{i=1}^{N} P_i^2,$$

where N is the total number of crops and P_i represents acreage proportion of the i-th crop in total cropped area. Its value is bounded by 0 and 1. With the increase in diversification, the HI would decrease. The index value is 1 when there is a complete specialization and approaches 0 as N gets large, i.e. if diversification is perfect. The major limitation is that it cannot assume the theoretical minimum, i.e. 0 for smaller values of N (number of crops). HI is a measure of concentration, the transformed value $(1-\mathrm{HI})$ is used instead of HI. For avoidance of confusion to compare with other indices, sometimes $(1-\mathrm{HI})$ is used instead of HI for the measure of diversification.

Simpson Index (SI) SI is given by the following formula:

$$\mathrm{SI} = 1 - \sum_{i=1}^{N} P_i^2,$$

where N is the total number of crops and P_i represents acreage proportion of the i-th crop. Its value also lies between 0 and 1. SI is a measure of diversification and is measured as $\mathrm{SI} = 1 - \mathrm{HI}$. This SI is most widely used instead of HI for the measure of diversification.

Ogive Index (OI) OI measures deviation from the benchmark given by equal proportion of each crop. If there are N crops, the norm used for measuring deviations is $1/N$. The formula for OI is given below:

$$\mathrm{OI} = \frac{\sum_{i=1}^{N} \left\{ P_i - \left(\frac{1}{N}\right) \right\}}{\left(\frac{1}{N}\right)}.$$

The OI is also a measure of concentration. Similarly, as in the HI, the transformed value $(1 - \mathrm{OI})$ is used more often. The major limitation of this index is that upper bound tends to approach 0 in case of perfect concentration, i.e. $N \rightarrow 1$, $P_i \rightarrow 1$, and $1/N \rightarrow 1$. So, it implies that the index approach 0 in extreme cases of perfect concentration as well as perfect diversification.

Entropy Index (EI) Entropy measure is regarded as an inverse measure of concentration having logarithmic character. Entropy is calculated by the formula:

$$\mathrm{EI} = -\sum_{i=1}^{N} P_i \log P_i$$

Or

$$\mathrm{EI} = \sum_{i=1}^{N} P_i \log\left(\frac{1}{P_i}\right).$$

The index would increase with the increase in diversification and it approaches 0 when there is perfect concentration, i.e. when $P_i = 1$. The upper bound of the index is $\log N$. The upper limit of the entropy index is determined by the base chosen for taking logarithm and the number of crops. The upper value of the index exceeds 1, when the number of crops is higher than the value of the logarithm's base and its value is less than 1 when the number of total crops is less than the value of the logarithm's base. Hence, the major limitation of EI is that it does not give standard scale for assessing the extent of diversification.

Modified Entropy Index (MEI) MEI is used to overcome the limitation of EI by using variable base of logarithm instead of fixed base of logarithm. It is computed by the formula given below:

$$\mathrm{MEI} = -\sum_{i=1}^{N} P_i \log_N P_i$$

Or

$$\mathrm{MEI} = \frac{\mathrm{EI}}{\log N}.$$

Here the important fact is that logarithm base is shifted to N number of crops. This index too is bounded by the number 0 and 1. The index assumes lower limit 0 when there is complete concentration, and upper limit of 1 in case of perfect dispersion.

The MEI gives uniformity and fixity to the scale used as norm to examine the degree of diversification. The major limitation of this index is that it measures the deviations from equal distribution among existing number of crops only, and does not incorporate the number of crops in it. This index measures diversification given the number of crops and the index is not sensitive to the changes in the number of crops.

Composite Entropy Index (CEI) This index has all the desirable properties of MEI. CEI is used to compare diversification across situations having different and large number of activities since it gives due weight to the number of crops. The formula of CEI is as follows:

$$\mathrm{CEI} = -\left(\sum_{i=1}^{N} P_i \log_N P_i\right) \cdot \left(1 - \frac{1}{N}\right)$$

Or

$$\mathrm{CEI} = \mathrm{MEI} \cdot \left(1 - \frac{1}{N}\right).$$

The CEI has two components, viz., distribution and number of crops, or diversity. The value of CEI increases with the decrease of concentration and rises with the number of crops. Both the components are bounded by 0 and 1. Since the index uses $-\log N\, Pi$ as weights, it assigns more weights to lower quantity and less weight to higher quantity.

12.3.2 Constraint Analysis

There is no single best method for identifying client constraints. A review of the available literature regarding the socioeconomic development of the region helps to identify some important constraints. Unfortunately, this information is rarely synthesized in a usable form. Therefore, the most important step in identifying client constraints is to review existing sources of information and gather fresh information from the clients as well as concerned officials and key informants living in the village or having knowledge about the village, its resources, problems, programmes, and plans. In this study, the same was done through rapid rural appraisal (RRA) and FGD in all the four villages. Based on the literature survey, initially, we identified an exhaustive list of constraints encompassing socioeconomic, technological, institutional, and policy constraints. The list was then shortlisted (and revised) through interactive process with key informants consulted RRA and FGDs. Subsequently, the constraints having social and economic significance were shortlisted for which information was gathered from each household. Accordingly, we finalized with 12 constraints for crop diversification for interrogation of the sampled households. The constraints are measured in a five-point scale for their severity with a maximum (minimum) value of '5' ('1') when the constraint is extremely severe (negligible). The severity of each constraint was assessed through the concerned household's perception based on his/her own experience.

12.4 Findings of the Study

12.4.1 Extent of Crop Diversification

The extent of crop diversification is analysed from two points of view: first, by examining the number of crops grown by different livelihood groups, and second, by constructing different types of diversification indices.

So far as the number of crops is concerned, on an average, more than three crops are grown by the landholding class in Burdwan district. The average highest numbers of crops are grown by cultivator group, followed by salaried group in this district (Table 12.1). While, in Purulia the highest number of crops grown by a household is only 2. In Purulia, as high as 41 % of sample households grows only one crop and rest of the households grows two crops per annum. In fact, there is not a single household, among sample households, who grew more than three crops in a year. The low level of crop diversification in Purulia may be due to distress induced by the agro-climatic factors, particularly due to erratic rainfall pattern and lack of any kind of irrigation facilities in the study villages.

The values of HI, SI, OI, EI, MEI, and CEI are shown in Table 12.2 and Table 12.3. It is evident from the above tables that the level of diversification is very low for all the livelihood groups in Purulia perhaps because of agro-climatic and socioeconomic constraints. The level of crop diversification, as measured through the various indices, is of moderate magnitude for all the livelihood groups in Burdwan. The magnitude is highest for cultivators followed by salaried class. From the above analysis, one thing is clearly emerging that irrespective of tools or methods used, the estimated level of crop diversification is much higher in Burdwan district than that in Purulia; and among different livelihood groups, salaried class and cultivators are in a better position perhaps due to their asset base.

Table 12.1 Average number of crops grown in Burdwan and Purulia district

Livelihood groups	Average no. of crops grown	
	Burdwan	Purulia
Agriculture and allied activities	4.36	2.00
Agricultural labourer	1.88	N.A.
Non-agricultural labourer	2.77	1.74
Salaried group	4.00	1.80
Casual labourer	2.80	1.00
Petty business	3.00	2.00
Others	3.20	1.80
All occupations	3.35	1.79

Table 12.2 Crop diversification indices in Burdwan district

Livelihood groups	Burdwan					
	HI	SI	OI	EI	MEI	CEI
Agriculture and allied activities	0.3977	0.6023	2.9775	1.1200	0.4864	1.0080
Agricultural labourer	0.6650	0.3350	5.6498	0.4967	0.2157	0.4470
Non-agricultural labourer	0.5312	0.4688	4.3119	0.7997	0.3473	0.7198
Salaried group	0.4118	0.5882	3.1183	1.0669	0.4633	0.8180
Casual labourer	0.6263	0.3737	5.2628	0.6578	0.2857	0.5920
Petty business	0.5435	0.4565	4.4350	0.8514	0.3698	0.7663
Others	0.5172	0.4828	4.1172	0.8637	0.3751	0.7774
All occupations	0.5056	0.4944	4.0313	0.8782	0.3814	0.7800

Table 12.3 Crop diversification indices in Purulia district

Livelihood groups	Purulia					
	HI	SI	OI	EI	MEI	CEI
Agriculture and allied activities	0.9554	0.0446	8.5540	0.1087	0.0472	0.0978
Agricultural labourer	NA	NA	NA	NA	NA	NA
Non-agricultural labourer	0.9600	0.0400	8.5999	0.0900	0.0391	0.0810
Salaried group	0.9458	0.0542	8.6576	0.0781	0.0339	0.0703
Casual labourer	1.0000	0.0000	9.0000	0.0000	0.0000	0.0000
Petty business	0.9376	0.0624	8.3760	0.1425	0.0619	0.1283
Others	0.9462	0.0538	8.4622	0.1202	0.0522	0.1081
All occupations	0.9574	0.0426	8.5933	0.0926	0.0402	0.0834

12.4.2 Nature of Crop Diversification

Table 12.4 shows the major changes in cropping patterns in the study area between 1990–1995 and between 2000–2005. In Burdwan district, spectacular change is observed in boro cultivation. Within the last 10–15 years, there has been a fourfold increase in boro rice cultivation in the study areas of Burdwan district due to rapid development in irrigation facilities. Irrigation coverage under both minor and canal irrigation has increased. A perusal of the table indicates that the cropping pattern in Purulia is mostly stagnant with monocropping of kharif rice. While in Burdwan, substantial crop diversification took place towards boro rice, potato, oilseeds, and vegetables away from sugarcane and wheat. Earlier, in the study villages of Burdwan district, potato was cultivated mainly for self-consumption. However, at present potato is cultivated mainly for the market as a cash crop. The other point to note is that the area under pulses is negligible. Production of pulses requires no such irrigation facilities. However, as the irrigation facilities are available, the farmers of Burdwan district showed preference towards the cultivation of paddy and potato. However, in present situation pulses are remunerative crops as the prices of these crops are substantially higher than other crops and require less water. So, farmers must be guided in this direction by suitable way. A notable feature of agriculture in Purulia is that pesticide and chemical fertilizer uses are very limited. The problem of crop disease is not much serious. The main problem of this area is the lack of irrigation facilities. Cultivation is totally rainfed. Farmers are willing to go for multiple cropping, but they cannot do this because of lack of irrigation facilities and erratic rainfall. So farmers in Purulia, where water scarcity is severe, should be encouraged to undertake crop diversification towards pulses, oilseeds, and horticultural crops like pumpkin, papaya, guava, etc., the cultivation of which requires less water.

12.4.3 Constraints to Crop Diversification in Burdwan District

The major constraints to crop diversification in Burdwan district for different livelihood groups are shown in Table 12.5. As evident from the table, the main constraints are high price instability and low prices of agricultural goods, pest-diseases-insect problem, lack of awareness and training, non-availability of seeds, poor asset base, lack of marketing facilities, and lack of credit facilities.

Low Prices of Agricultural Goods Prices of agricultural goods are the most important constraint to crop diversification in Burdwan. In recent years, the cost of production has increased substantially, but the farm gate prices of the agricultural goods have not increased to that proportion. If there is an increase in prices for some crops then the profits accrue to the middleman, the farmers do not get the benefits. As a result the profitability of crop cultivation has become low, e.g. potato cultivation. Previously, potato was cultivated only for home consumption. However, at present potato is cultivated for market also. At the time of harvest, prices of potato remain

Table 12.4 Cropping pattern in the study villages of Burdwan and Purulia district (in percentage)

Season	Crop	Burdwan		Purulia	
		1990–1995	2005–2010	1990–1995	2005–2010
Kharif	Paddy	79.18	50.32	97.53	97.75
	Sugar cane	2.07	0.49	0.05	0.00
Rabi	Mustard	5.17	8.95	0.00	0.00
	Til	2.07	4.44	0.00	0.00
	Potato	3.22	5.50	0.15	0.00
	Pulses	0.01	0.05	0.12	0.00
	Wheat	0.79	0.00	0.18	0.00
	Aurum	0.05	0.02	0.00	0.00
	Vegetables	0.00	1.00	1.97	2.25
Boro	Paddy	7.44	29.23	0.00	0.00

Table 12.5 Rank of some major constraints to livelihood diversification in Burdwan district

Constraints	Score	Rank	Most vulnerable groups
Asset/capital	3.99	I	Petty business, others
Credit problems	3.75	II	Others, non-agricultural labourer
Lack of awareness and training	3.73	III	Casual labourer, petty business
Fear of taking risk	3.41	IV	Petty business, casual labourer
Lack of infrastructure	3.24	V	Petty business, casual labourer

very low. Without adequate storage facilities, farmers are forced to sell their crop at low prices. Sometimes they were not even able to recover the production cost.

Pest-Diseases-Insect Problem Pest-diseases-insect problem is one of the serious constraints to crop diversification in Burdwan. There is an emergence of new pests and diseases in the district. Several insect pests also developed resistance to pesticides. It is a well-known fact that high yielding variety (HYV) seeds are easily susceptible to pest-diseases-insect problem. To manage this problem high quality pesticides are required. However, most of the farmers are resource poor and are not able to buy the high quality pesticides, which are costlier. Therefore, they use low quality pesticides yielding low productivity. Some farmers even reported that the pesticides available in the local market are duplicate pesticides of branded companies and thus failed to given result even after their application.

Lack of Awareness and Training Modern agricultural technology is continuously improving and changing. Most of the farmers are unaware of the improved agricultural technology. Extension services provided by the state are not reaching to the poor farmers. The farmers of this region have demanded for seminar, workshop, and symposium to make them aware about the current technology. They are also interested to cultivate high-value crops, but they do not get the necessary information and training from the concerned departments. Dissemination of technology regarding integrated pest management, bio-pesticides, pest and vector management, etc. is highly required.

Non-Availability of Seeds Non-availability of good quality seeds and their exorbitant prices are the acute problems of the study area. In the past, the farmers used to preserve seeds domestically. At present, almost all farmers use HYV seeds. Though they preserve some amount of seeds, they have to purchase a good proportion every year. The problem is that supply of good quality seeds are not enough. According to the farmers of this area, seed supplied by the government agency are of poor quality though their prices are also quite low. Most of the farmers purchased seeds from private agencies at high prices. Many farmers complained that the village seed dealers often sold duplicate seeds. Some farmers are interested to diversify their cropping through high-value crops. However, they cannot do this because of non-availability of quality seeds.

Poor Asset Base A majority of the sample households reported that lack of capital is one of the important constraints to crop diversification. Most of the farmers in this region are resource poor. However, modern technologies are capital intensive. Therefore, farmers cannot adopt all the improve technologies due to their poor resource base.

Lack of Marketing Facilities Farmers face problems to sell their products. There is no regulated market or co-operative society where they can sell their products at reasonable prices. They are forced to sell their products to the local agents at low prices under distress.

Lack of Credit Facilities To get the required credit (institutional), farmers have to go through complicated paper work and processes. There is mortgage restriction, which is in general holding of land acreage. There are two main problems farmers face in getting institutional loan. Firstly, average farmers have landholding less than the requirement for institutional loan. Secondly, if they have the required land holding they do not possess the proper record for the same. As a last resort, they are forced to borrow from the moneylender at an exorbitant rate of interest (24–60 % per annum).

Earlier, i.e. before 10–15 years, the main constraints were lack of awareness or training, and lack of irrigation facilities (Khatun and Roy 2012). During this period, considerable improvement took place both in terms of expansion of irrigated areas and awareness about modern farm practices. Irrigation coverage has increased due to both minor irrigation and canal irrigation. At present, a new problem arises regarding irrigation. Indiscriminate use of ground water is causing fall of water table. Every year the water table decreases at an alarming rate. The farmers have to deepen their well (shallow pumps) regularly due to declining water table depth. It causes a serious ecological imbalance and scarcity of drinking water. As a result, cost of irrigation has also increased.

12.4.4 Constraints to Crop Diversification in Purulia district

In Purulia district, the main constraints to crop diversification are quite different from Burdwan district. The major constraint to crop diversification in Purulia

Table 12.6 Rank of some major constraints to livelihood diversification in Purulia district

Constraints	Score	Rank	Most vulnerable groups
Road and transport problems	4.10	I	Petty business, agriculture and allied activities
Asset/capital problems	3.61	II	Agriculture and allied activities, non-agricultural labourer
Credit problems	3.53	III	Agriculture and allied activities, non-agricultural labourer
Lack of awareness and training	3.45	IV	Agriculture and allied activities, non-agricultural labourer
Lack of infrastructure	3.37	V	Agriculture and allied activities, others

district are lack of irrigation facility, recurring drought, erratic rainfall, poor asset base, lack of awareness and training, and lack of credit facilities (Table 12.6).

Lack of Irrigation Facility The level of irrigation in Purulia district is very low. In fact, in the sample villages there were virtually no irrigation facilities. Cultivation is fully rainfed. As a result, farmers are unable to diversify their cropping pattern despite their immense interest to do so.

Drought Drought is very common in Purulia. The problems faced due to drought situation are mainly loss of agricultural production due to dependence on rainfed agriculture. Drought also affected the livestock rearing due to the non-availability of grass and fodder. All these lead to the loss of access to food. Drought is a recurring problem in this region.

Erratic Rainfall The amount of rainfall in the district is not only low but also very irregular. As irrigation infrastructure has not developed, erratic rainfall causes severe problems to cultivator households. Cultivation has become the gamble of monsoons.

Poor Asset Base Lack of capital is an important constraint for crop diversification. The resource-poor farmers are not capable of making any investment in agriculture. As a result, productivity of agriculture is very low in Purulia district. Most of the farmers are not even capable of purchasing the fertilizers required for cultivation.

Lack of Awareness and Training Most of the farmers in Purulia district are small and marginal farmers and illiterate. They are not aware of the modern techniques of cultivation. There are no extension services provided by government or any NGO to make the farmers aware of modern farming practices.

Lack of Credit Facilities Availability of credit facilities is important for cultivation particularly to the resource-poor households. However, without any access to institutional credit, farmers are forced to go to the moneylender. They are forced to borrow at a high rate of interest from the moneylender.

To sum up, the principal constraints faced by the rural households in the study area are of various kinds. While most of them are socioeconomic in nature, some constraints are of agro-ecological nature. There are few constraints, which are technical or institutional in nature like non-availability of quality seeds resistant to extreme climates, and pests and diseases, etc. Many of the institutional constrains have their origin in policy failure. Rural households in West Bengal, however, do not face identical constraints. Spatial variation leads to cross sectional heterogeneity thereby influencing diversification pattern. Property rights in productive assets such as land and livestock, labour availability, and access to credit differs across livelihood groups. Therefore, though all the livelihood groups face these constraints because of poor asset base, the severity of the constraints are more for the landless labourer groups and least for the resource-rich salaried class.

High volatility in prices, absence of market and lack of access to technical knowhow are the main constraints faced by the cultivators group. Cultivators are faced with price volatility of outputs particularly for potato and other vegetables. Cultivators are more susceptible to such crisis. At present, no institutional arrangement exists, except few cold storages for potato, to protect the poor farmers from price volatility. Price stability can be ensured through better market and storage facilities. In (2010–2011) the state government has taken special initiatives to procure potato from the farmers and export the same in South Asian markets. Such initiatives helped the potato growers in Burdwan and Hoogly, which are yet to spread to the remote areas and for other vegetables. A large proportion of small and marginal farmers gain livelihoods through production on small pieces of land. For these households, timely availability or access to credit and improved methods of production are quite critical for their livelihood and for crop diversification. It is also found out that cropping pattern in Purulia is mostly stagnant with monocropping of kharif rice, where as in Burdwan, substantial crop diversification took place towards boro rice, potato, oilseeds, and vegetables away from sugarcane and wheat.

12.5 Conclusion and Policy Implications

Crop diversification is an important strategy for small and marginal farmers to augment income and enhance employment. West Bengal has been diversifying at a faster rate than the all India level. However, this has occurred only in some selected areas, not in all over the state. Level of diversification varies across districts and livelihood groups. This study analysed the nature and extent of crop diversification in West Bengal across livelihood groups in two districts utilizing various types of diversification indices. The results show that irrespective of tools or methods used, the estimated level of crop diversification is much higher in Burdwan district than Purulia and among different livelihood groups, salaried class and cultivators are in a better position. Constraints for crop diversification in study area are identified in a systematic way. High volatility in prices, absence of market, and lack of access to technical knowledge are the main constraints for crop diversification in Burdwan district. The major constraint to crop diversification in Purulia district are lack of irrigation facility, recurring drought, erratic rainfall, poor asset base, lack of awareness and training, and lack of credit facilities. To eradicate these constraints and to pave the way for successful crop diversification, some government interventions are needed. Based on the findings of this study, following interventions are suggested:

1. Drought proofing should be accorded high priority over drought relief particularly in drought prone areas like Purulia. Skewed policies have led to the decay of traditional water harvesting structures in the state. The state has a network of more than 25,000 ponds, which can be used successfully to combat drought. However, poor maintenance and sheer neglect have resulted in many of these ponds being ineffective. Efforts should be taken to dig new ponds and/or renovate the existing ponds under NREGS (National Rural Employment Guarantee Scheme) work.
2. Since lack of credit facilities and poor access to institutional credit is acknowledged as one of the most important constraints inhibiting crop diversification, the rural financial systems need to be revamped. The microfinance revolution, as observed in Bangladesh can be a model in this regard but the replicability and efficacy of such a model is still uncertain.
3. Extension system must be strengthened so that farmers can be aware of modern farm practices.
4. Crop insurance should be implemented with more priority, which can protect the farmers from adverse shocks.
5. In Purulia district, cultivators should be motivated to cultivate the crops like pulses and oilseeds and horticultural crops like pumpkin, papaya, guava, etc. that require less water.

References

Bhalla GS, Singh G (1997) Recent development in Indian agriculture: a state level analysis. Econ Political Wkly 32(13):A8–A29

De UK, Chattopadhyay M (2010) Crop diversification by peasants and role of infrastructure: evidence from West Bengal. J Dev Agric Econ 2(10):340–350

Ghosh BK (2010) Essence of crop diversification: a study of West Bengal agriculture during 1970–1971–2004–2005. J Dev Agric Econ 2(11):368–381

Joshi PK, Gulati A, Birthal PS, Tewari L (2004) Agricultural diversification in South Asia: patterns, determinants and policy implications. Econ Political Wkly 39(24):2457–2468

Joshi PK, Gulati A, Birthal PS, Rao PP (2005) Agricultural diversification and vertical integration in India: will smallholders participate? MITD, International Food Policy Research Institute, Washington, D.C. (Mimeo)

Khatun D, Roy BC (2012) Rural livelihood diversification in West Bengal. LAP LAMBERT Academic, Germany

Kumar P, Mruthyunjaya, Birthal PS (2003) Changing consumption pattern in South Asia. Paper presented in the international workshop on agricultural diversification and vertical integration in South Asia organized by FICCI-ICRISATIFPRI on 5–6 November 2003 in New Delhi

Vyas VS (1996) Diversification in agriculture: concept, rationale and approaches. Indian J Agric Econ 51(4):636–643

Chapter 13
Problems of Crop Diversification in West Bengal

Debisree Banerjee and Uttam Kumar Bhattacharya

13.1 Crop Diversity and Its Role in Agriculture

Agriculture is one of the most important economic activities in most of the developing countries including India. The agricultural sector in India together with allied activities contributed about 14 % of the gross domestic product (GDP) of which 12 % was contributed by agriculture alone during 2011–2012, at constant price 2004–2005. Agriculture also provides about 58 % of total employment. However, there is a decline in contribution of agriculture towards national growth over time. This is evident from the fact that where the overall GDP has grown up by an average of 6.2 % during 2007–2008 to 2011–2012, agricultural sector GDP increased by only 3.6 % during the same period (Economic Survey 2012–2013). Still, agriculture remains an important sector in the context of food security (Bhalla 2006).

In order to increase agricultural productivity and employment, adaptation of crop diversity can be an important method. This would enable to use a variety of characteristics of plants in agriculture. Crop diversity helps to increase productivity and to optimise the utilization of scarce land resource along with increase in employment opportunities. However, various uncertainties in Indian agriculture, particularly related to introduction of new technology is not accompanied by adequate measures for natural resource conservation, which has led to degradation of land. With increase in irrigation facility, over-dependence and over-exploitation of ground water caused a fall in the water level. The reduced availability of water increases possibility of occurrence and severity of drought. Thus, crop production dependent upon rainfall, becomes quite vulnerable to the effects of drought (Ghosh 2008). Mohan (2006) pointed out that a silent transformation is taking place in

U. K. Bhattacharya (✉) · D. Banerjee
Institute of Development Studies Kolkata (IDSK), 27/D Block DD, Sector-1, Saltlake, Kolkata 700064, West Bengal, India
e-mail: uttam.bhattacharya@gmail.com

D. Banerjee
e-mail: debisreeb3@gmail.com

M. Ghosh et al. (eds.), *Diversification of Agriculture in Eastern India*,
India Studies in Business and Economics, DOI 10.1007/978-81-322-1997-2_13

rural areas calling for diversification in agricultural production and value addition processes in order to protect employment and incomes of rural population. In the changed scenario, strong and viable agricultural financial institutions are needed to cater to the requirements of finance for building the necessary institutional and marketing infrastructure.

In this chapter, we have tried to identify the problems of crop diversification in West Bengal. We have studied the features of agriculture in the state and explained how the rural credit system and agricultural insurance could play a significant role in exploring and sustaining the benefits of crop diversification in West Bengal.

13.2 Agriculture in West Bengal

West Bengal is one among the most important agricultural states in India. It is the sixth largest contributor to the country's GDP, which is attributed to agriculture and the allied sector. Its portion of agricultural land is amongst the highest of all Indian states (Economic Survey of India 2011–2012). The state is divided into six agro-climatic zones (hilly region, terai region, old alluvial region, new alluvial region, coastal region and western laterite region), which brings considerable diversity in its agricultural activities. The agricultural sector of the state is characterized by the predominance of numerous small and marginal farmers tilling major portion of the total operated area of the state (Table 13.1). The average size of land holding is < 1 Ha. Since the scope for bringing more area under cultivation is very limited so greater emphasis has been laid mainly on increasing the productivity of different crops.

13.3 Crop Diversification in West Bengal

The diversity in the six agro-climatic zones resulted in inter-regional differences. In West Bengal, the extent of crop diversity has been mainly attributed to tenancy reforms and expansion of new technology both at state and district level. Crop diversity can be mainly understood from the cropping intensity, which is quite favourable in the case of West Bengal. According to the *Economic Review of West Bengal 2012-2013,* the state's cropping intensity was 180 % during 2011–2012 compared to 177 % in 2010–2011. The gross cropped area (GCA) of the state also increased by 6 % during this period.

The West Bengal State Development Report 2010 attributed the change in crop intensity primarily to multiple cropping supported by irrigation, which in turn help in increasing the GCA. The report further pointed out that the increase in cropping intensity of the state in 1990s mainly occurred due to increase in cropped area under *Boro* rice. The share of cropped area under the traditional variety of *Aman* rice declined sharply from < 58 % in 1980 to 47 % in 2000 but increased marginally thereafter. The acreage shared by *Boro* (the HYV) on the other hand increased from less than 5 % to around 18 % during the same period. Ramchandran

Table 13.1 Percentage distribution of area of operational landholding among different classes, 2005–2006. (Source: Agricultural census of India 2005–2006)

Size of holding	Percentage of holding	
	Number	Area
Marginal (<0.5–1.0 ha)	81.17	52.80
Small (1.0–2.0 ha)	14.38	30.09
Semi-medium (2.0–4.0 ha)	4.04	14.55
Medium (4.0–10.0 ha)	0.4	2055
Large (>10.0 ha)	0.01	0.02

et al.[1] highlighted that West Bengal's rural economy was characterized by rapid growth in the 1980s and early 1990s. The major features of growth, which was particularly marked in the rice economy of the state, were rapid growth in aggregate production; growth in yields per hectare, particularly in the *Boro* (*or rabi*) season, and an overall narrowing of the gap between districts with respect to production and yield performance.

Apart from the increase in cultivation of *Boro* rice, some other factors were also considered significant for increasing in crop diversity of West Bengal. The percentage increase of acreage of oilseeds, particularly mustard, was nearly doubled during 1980–2006. The area under potato also increased magnificently during the same period whereas share of cropped area for jute declined for this period. Agriculture in West Bengal has also been diversifying towards high value crops such as fruits, vegetables and flowers. It is one of the largest producers of fruits and vegetables in the country. Thus, crop diversification in West Bengal has become quite prominent. However, the cultivation of various high yielding varieties of crops, fruits and vegetables often become victim of production, marketing and other risks.

13.4 Agriculture and the Risk in West Bengal

Considering the fact that cropping diversity can reduce several limitations and risks involved in traditional methods of agriculture, it cannot eliminate the risk completely. Moreover, adaptation of new technologies in modernized agriculture such as use of genetically modified crops increases producer liability risk. Agricultural productivity largely depends upon farm size, which is considerably low in West Bengal compared to that of other agricultural states of India.

The marginal landholders increased from 61 % in 1970–1971 to nearly 89 % in 2002–2003. Further maximum portion of the landholding are concentrated among the poor and vulnerable section. The level of farmers' indebtedness explains the situation. According to the *State Development Report 2010*, about 50 % of the farm households were reported to be indebted and 55 % of them were cultivators. More-

[1] http://www.agrarianstudies.org/UserFiles/File/6Ramachandran_et_al%20_Food_Security_and_Crop_Diversification...pdf, 04 Jan 2012.

Table 13.2 Incidence of indebtedness in West Bengal, 2012. (Source: Agricultural Statistics at a Glance 2012)

% of indebted farmer households				
Marginal (up to 1.0 ha)	Small (1.01–2 ha)	Semi-medium (2.01–4.00 ha)	Medium (4.01–10.00 ha)	Large (> 10.00 ha)
88.7	8.5	2.4	0.4	0

over, the incidence was highest among the marginal farmers (Table 13.2). Most of the farmers in West Bengal have appeared to be prone to several kinds of risks.

Sainath (2011)[2] noted that West Bengal is in a very fragile situation and in no way isolated from the general agrarian crisis. He pointed to the vast land used for cultivation of tea, a cash crop that is going to be at the mercy of global prices. Floods, landslides, cyclone, droughts and occasional earthquakes have frequently affected the state. More than 50% of the areas of West Bengal are susceptible to floods. It has been one of the most cyclone-affected territories on one hand and on the other hand, droughts are common for many of the districts such as Birbhum, Purulia and Bankura[3].

In fact, in order to sustain the benefits from crop diversity technological and institutional supports could be considered very necessary. Here, technological support would mean using modern tools and institutional support would mean assistance from banking and insurance sectors. Modern agriculture with increased need for modern inputs has also increased the demand for agricultural credit. Farmers' indebtedness has been a serious concern for India as well as for the state. As stated earlier majority of farmers are indebted over here and they have to borrow regularly from different sources to continue with their farming activities.

The Situation Assessment Survey of farmers on indebtedness of farmer households (2003) NSS 59th Round, reported that in West Bengal on an average, if ₹ 1000 were lent to farmers, then the share of banks were ₹ 285, followed by 192 contributed by co-operative banks, friends and relatives contributing ₹ 154 and ₹ 130 by the agricultural money lenders. In addition, a major portion of the loan taken by the farmers was mainly used for non-farming activities. Further, agricultural credit is also subject to different challenges. For example, seasonal nature of production, dependence on biological processes, dependence on scarce natural resources with several natural and man-made risks. Here comes the role of insurance in agriculture.

[2] See Sainath; The Hindu, *Farmers suicide is not the crisis, it is the outcome*; 08 March 2011.

[3] See Nath et al. (2008) for details.

Table 13.3 Disbursement of credit to the agricultural sector, 2009–2010 to 2012–2013. (Source: http://www.slbcbengal.com accessed on 15 January 2014)

Year	Target (₹ in crore)	Achievement (₹ in crore)	Percentage of target
2009–2010	9015	8178	91
2010–2011	16,000	11,555	72
2011–2012	20,300	14,387	71
2012–2013	23,500	15,935	68

13.5 Role of Agricultural Insurance in West Bengal

It is often considered that insurance is an effective tool for reducing and/or eliminating risk, by which losses suffered by a few could be overcome through the contributions made by a similar group. According to Wenner (2005) agriculture is a risky business subject to price, climatic, geological and biological shocks. To deal with these kinds of risks coping strategies and financial management instruments are required. Insurance could be beneficial in improving access to credit by serving as a guarantee against involuntary default.

The basic object of insurance is to protect the insurer against the risks insured. A pool is created through contributions collected from the person, who is willing to insure in the form of premium by the insurance companies to protect from some kind of common risk. It is expected that this would serve the economy better in the best interest of the community and ensure that concentration of wealth does not take place to a common detriment. Agricultural insurance could open an avenue to cover the farmers for any loss occurring in agriculture, which covers wide objectives. It comprises of various kinds of insurance apart from different crop insurance schemes that protects farmers in case of crop failure due to natural and manmade hazards, perils and risks due to different economic policies pursued by the government.

13.6 Scope of Agricultural Insurance and Crop Diversity

As stated earlier, change in cropping pattern intensifies the need for stronger financial institutions. Owing to the prevalence of small and marginal farmers', the dependence on credit for agricultural activities has been growing over the years in West Bengal. Disbursement of credit to agricultural sector almost doubled in between 2009–2010 and 2012–2013 (Table 13.3).

Cropping pattern varies widely in West Bengal with *Aman* variety produced almost everywhere as the leading crop. The other major crops grown over here are *Boro rice, Aus rice, potato, pulses, wheat, jute and oilseeds*. Considering the situation agricultural insurance (National Agricultural Insurance Scheme) is mainly

Table 13.4 Details of different crop insurance schemes adapted in India. (Source: AICL various years)

Schemes	Year	Features	Crops covered
First individual approach	1972–1978	Voluntary, implemented in 6 states	H4 cotton, groundnut, wheat, potato
Pilot crop insurance scheme	1979–1984	Only for loanee farmers, voluntary, implemented in 13 states	Food crops, oil seeds, cotton, potato, cereals, millets and pulses
Comprehensive crop insurance scheme	1985–1999	Compulsory for loanee farmers, implemented in 16 states and 2 union territories	Food crops, oil seeds, cotton, potato, cereals, millets and pulses
Experimental crop insurance scheme	1997–1998	For small and marginal non-loanee farmers, implemented in 5 states	Food crops, oil seeds, cotton, potato, cereals, millets and pulses
National agricultural insurance scheme	1999 and continues	Compulsory for loanee and voluntary for non-loanee farmers, implemented in all the states	All food grains, oil seeds, and annual orticultural/commercial crops

Table 13.5 Performance of the NAIS in West Bengal from rabi 1999–2000 to kharif 2012. (Sources: AICL and Economic Review of West Bengal 2011–2012)

No. of farmers covered (in '000)	Area insured (in ha) as percentage of GCA	Sum insured (₹ in crore)	Gross premium (₹ in crore)	Claim (₹ in crore)	No. of farmers benefited (in '000)
10,835	62.20	11111.13	465.71	955.36	2,268

meant for *Aman, Aus and Boro rice, wheat, mustard, gram, jute, potato, mung, masoor, arhar, til and linseed.* Crop insurance started in 1972 in West Bengal along with other parts of India with introduction of 'First Individual Approach Scheme'. However, with several improvements and modifications over time and depending on the situational requirements, it could be enforced on a large scale only with implementation of the NAIS during *rabi* 2000–2001 (Table 13.4).

The scheme was a joint effort of the central and state governments for the well-being of the farmers. Agricultural Insurance Company of India Ltd. (AICL) revealed that from *rabi* 2000–2001 to *kharif* 2012, more than 108 lakh farmers were covered under the NAIS with a gross premium of about ₹ 465 crore only in West Bengal (Table 13.5). The said scheme was available to both borrowers as well as non-borrowers. It was a compulsory scheme for the borrowers and voluntary for the non-borrowers. It could cover all food grains, oilseeds and annual horticultural/commercial crops. The scheme started on the basis of both the 'area approach' (for widespread calamities) and the 'individual approach' (for localized calamities such as hailstorms, landslides, cyclones and floods) in all the states and union territories.

Table 13.6 Seasonal disparity in coverage of the NAIS from rabi 2000–2001 to rabi 2010–2011 in West Bengal. (Source: AICL, West Bengal 2012–2013)

Seasons	No. of farmers covered (in '000)	Area insured (in '000 ha)	Sum insured (₹ in crore)	Gross premium (₹ in crore)
Kharif	5136	2777.00	3922.00	97.00
Rabi	5699	2716.89	7189.29	368.52
Total	10835	5494.89	11111.13	465.71

In West Bengal we have noted, predominance of small and marginal farmers with highly erratic income. They need to depend upon various formal and informal sources of loan continuously for cultivation. Even the small amount of premium is considered to be a burden for them. Moreover, the complications involved in the scheme and delays in repayment mechanism have reduced the faith of the farmers on the scheme and made it dependent on the institutional credit system. Moreover, there is a problem of adverse selection[4]. In West Bengal; the major crops cultivated are *Aman* and *Boro* paddy. *Aman* is a rainfed crop cultivated during *kharif* season, whereas *Boro* is a *rabi* crop for which risk associated is higher due to the irrigation problem. Due to this problem, seasonal disparity in coverage could be observed (Table 13.6).

The NAIS has been the only agricultural insurance scheme operating in India on a large scale and providing considerable benefits. However, we have observed certain limitations of the scheme for which it was not able to extend the desired results. Raju and Chand (2007 and 2009) mentioned that crop insurance in India still depends upon institutional credit system with very low voluntary participation from non-borrower farmers and lack of awareness even among the borrowing farmers, with high claims to premium ratio indicating loss of operation. Further, the problem of adverse selection and moral hazard is common with ineffective actuarial premium rates. The scheme is also often not preferred by the farmers as there are complications involved with it including delays in claim settlement. The Department of Agriculture and Cooperation of India pointed out some of the limitations in the Joint Group Report (2004). Those could be highlighted as follows:

- Threshold yield on which indemnities are calculated is moving average yield of preceding 3 years for rice and wheat and 5 years for other crops, multiplied by the level of indemnity. The method often does not provide adequate protection to farmers in their time of need.
- Indemnity limit (i.e., the limit applied on average yield to produce threshold yield) is 90, 80 and 60 % corresponding to low risk, medium risk and high-risk

[4] Adverse selection: It refers to a market process in which "bad" results occur when buyers and sellers have asymmetric information (i.e. access to different information). In insurance sector it means selective participation in crop insurance program. For example, participating during potential adverse seasons and avoiding the normal ones.

areas. The indemnity level of 60 % cannot adequately cover the risk of small and medium adversities.

- The scheme covers risk only from sowing to harvesting. There are certain pre-sowing and post-harvesting risks that should also be taken into consideration.
- There are problems of adverse selection. In *kharif* a farmers already get some indications of the monsoon before taking up the policy causing this problem.
- There is delay in settlement of claims as the claim processing starts only after the harvest of the crop. There is a time gap of 8–10 months between occurrence of loss and claim payment. Thus, the farmers do not receive benefit on time.
- The scheme is voluntary for non-loanee farmers with inadequate marketing. There is an evidence of lack of awareness among the farmers.

Considering the limitations of the NAIS, a Pilot Modified NAIS (MNAIS) has been implemented in 50 districts from *rabi* 2010–2011 including West Bengal. The major changes brought about were:

- Unit area of insurance reduced to village *Panchayat* levels for major crops.
- Indemnity for prevented/sowing/planting risk and for post-harvest losses due to cyclone included.
- Payment up to 25 % advance of likely claims as immediate relief.
- Minimum indemnity level rose to 70 % instead of 60 %.
- Actuarial premium will be paid for insuring crops and hence claims liability will be on the insurer.
- Uniform seasonality norms will be applicable for both *loanee* and non-*loanee* farmers.
- Private sector insurers provided with adequate infrastructure (ICICI Lombard, IFFCO-Tokio and Colamandalam-MS).[5]

In spite of modifications and re-modifications, we have noted several major limitations in the yield based NAIS and other crop insurance schemes. According to Swaminathan (2001), crop insurance though an efficient tool, credit linked to insurance was not successful. Hardly 4 % of fifteen million farmers of India took up crop insurance. Due to various limitations, there has been a slow convergence towards weather based insurance schemes. While crop insurance specifically indemnifies the cultivator against shortfall in crop yield, weather based crop insurance has become dependent on the weather conditions, which often affect crop production even when a cultivator took all the care to ensure good harvest. It has been also observed that weather insurance schemes also suffer from certain limitations. Those limitations could be in different forms such as, distance of the farmland from the weather station, non-coverage of perils other than mere weather calamities, shifts in climatic and weather patterns, difficulties in understanding formalities etc.

Nair (2010) pointed out that weather insurance could help in expanding the domain of crop insurance program in the country. She argued that it could be implemented in the context of horticultural crops for which age group wise yield estimates are

[5] Sinha (2004) explored the scope of private insurances in agriculture.

Table 13.7 Details of additional agricultural insurance schemes in West Bengal. (Source: AICL, West Bengal, 2010–2011)

Schemes	Year	Features
Farm income insurance	*Kharif* 2004	Covered yield losses suffered due to natural calamities, pests, diseases and market price fluctuations
Varsha Bima	*Kharif* 2006	Covered anticipated shortfall in crop yield on account of deficit rainfall
Potato insurance	*Rabi* 2006	Covered losses suffered by the insured in respect of cost of inputs due to happening of insured peril
Weather based crop insurance scheme	*Kharif* 2009	Covered losses due to adverse incidences of weather parameters like rain, frost, humidity etc.

unavailable. So according to Nair what was needed, is to address the challenges faced by both the area yield and weather insurance schemes; then the schemes could become complementary in providing efficient mechanism for transferring risks involved in agriculture. To improve the livelihood of farming community implementation of effective agricultural insurance schemes has become an urgent need.

13.7 Different Agricultural Insurance Schemes in West Bengal

The AICL has also introduced a few new schemes in West Bengal from time to time, considering their relevance in different parts of the state. Those were Weather Based Crop Insurance Scheme (WBCIS), Farm Income Insurance Scheme (FIIS), *Varsha Bima,* and Potato Insurance schemes (Table 13.7). Most of those schemes have suffered from different kinds of losses. So, those schemes were also forced to be stopped after experimental implementation for one or two seasons.

The FIIS was implemented in the districts of Birbhum, Jalpaiguri, North 24-Parganas and Murshidabad in *kharif* 2004. The scheme covered about 58 thousand farmers with gross premium collection of 120 lakhs. However, as the NAIS already covers yield risks and government already provides minimum support price so the scheme has become redundant and is not functioning at present.

Varsha Bima was applied on a pilot basis in *kharif* 2006 to cover anticipated shortfall in crop yield due to deficit rainfall covering 40 farmers and about 12 ha of land for a sum insured of ₹ 1.87 lakh against a premium of ₹ 9,218. The scheme reported a deficit of ₹ 0.45 lakh.

To provide financial assistance to potato growers of West Bengal potato insurance was introduced by Pepsico Company in rabi 2006 on pilot basis covering 24 potato farmers and 12 acres of land for a sum insured of ₹ 2.89 lakh against a premium of ₹ 22,463. Claim of ₹ 1.80 lakh was reported during the period and the scheme reported a deficit of ₹ 1.66 lakh (AICL 2011–2012).

Among those schemes, the WBCIS assess crop yield losses through measuring various weather parameters such as deficit and excess rainfall in *kharif* season and adverse incidences of weather parameters like frost, heat, relative humidity etc. during rabi season. It was implemented in West Bengal in *kharif* 2009 on pilot basis in nine blocks spreading over four districts of the state. It was applicable for non-*loanee* farmers as *loanee* farmers were already compulsorily covered by the NAIS. During this period about 2 lakh ha of crop area was affected due to '*Aila*'[6]. About 4 thousand farmers were covered, 2.2 thousand ha of land was insured and a total premium of ₹ 23 lakh only was collected against a sum insured of ₹. 432 lakh under the scheme. The AICL statistics noted hat during rabi 2008–2009, a claim of ₹ 10.52 lakh was paid in the districts of North 24-Parganas and East Midnapore. In North 24-Parganas alone, about ₹ 10.47 lakh was paid against 930.68 ha of land to cover the loss of crops due to excess rainfall during the '*Aila*' cyclone phase.

13.8 Summary and Concluding Observations

Agriculture in West Bengal has been going through a phase of transformation in post green revolution era. Here, most of the farmers appeared small and marginal with highly erratic income levels. Owing to the small landholding pattern there has been an increasing dependence on credit from different institutional and non-institutional sources. Emphasis was given on using modern methods and tools of cultivation in order to increase productivity. The plans for crop diversifications became leading schemes. Crop diversifications could create an avenue to open up new opportunities for higher income and help to overcome the limitations of small land holding. However, crop diversification cannot be sustained without technological aids and effective financial infrastructures. Here agricultural insurance schemes could help to form an important role.

Insurance in the field of agriculture helps to reduce several risks involved in farming. It is a dynamic concept, which has evolved over time with the changing farming needs. Earlier there was merely a yield-based scheme covering limited crops and limited farmers on individual basis. However, now it is much more than a yield based scheme. It could cover a large number of farmers and almost all kind of crops including the horticultural crops. What we need today is to explore the methods through which the benefits could reach to the needy ones.

[6] About 209,926 ha of crop area was affected due to '*Aila*'—Aila cyclone report 2009.

References

Agricultural Census (2001) www.agcensus.nic.in. Accessed 10 Dec 2013

Agricultural Statistics at a Glance (2010) www.agricoop.nic.in/Agristatistics. Accessed 10 Dec 2013

Agriculture Insurance Company of India Ltd (2010) www.aicofindia.org. Accessed 24 Dec 2013

Bhalla GS (2006) Condition of Indian peasantry. National Book Trust, New Delhi

Department of Economic Affair, Economic Division, Government of India, *Economic Survey*, New Delhi, Ministry of Finance, various issues

Ghosh M (2008) Economic reforms and indian economic development selected essays. Bookwell, New Delhi

Ministry of Agriculture, Department of Agriculture and Cooperation, Government of India (2004) Report of joint group on crop insurance, New Delhi, Ministry of Agriculture

Ministry of Finance, Government of West Bengal (2009–2010) Economic review, Ministry of Finance, Kolkata, various issues

Mohan R (2006) Agricultural credit in India: status, issues and future agenda. Econ Polit Wkly XLI(11):1013-1021 (March 18)

Nair R (2010) Crop insurance in India: changes and challenges. Econ Polit Wkly XLV(6):19–22 (Feb 6)

Nath SK, Roy D, Singh TK (2008) Disaster mitigation and management for West Bengal, India—an appraisal. General articles, April 10, pp 858–864

Planning Commission, Government of India, State Development Report (2009–2010). New Delhi Planning Commission, and New Delhi: Academy Press

Raju SS, Chand R (2007) Progress and problems in agricultural insurance. Econ Polit Wkly 42(21):1905–1908 (May 26)

Raju SS, Chand R (2009) Problems and prospects in agricultural insurance in India. NCAP Policy Brief, July

Sinha S (2004) Agriculture insurance in India scope for participation of private insurers. Econ Polit Wkly June 19, pp 2605-2612

Swaminathan MS (2001) Uncommon challenges and opportunities in Indian agriculture. Dec, pp 36–48

Wenner M (2005) Agricultural insurance revisited: new developments and perspectives in Latin America and the Caribbean. Rural Development Unit Sustainable Development Department, Inter-American Development Bank, New York

Chapter 14
Factors Influencing the Extent of Diversification in West Bengal

Debajit Roy

14.1 Introduction

Experts of the World Bank have suggested the package of measures to the problems affecting the agrarian economy of India, based on traditional food grains and as a way-out from nonviable and noncompetitive state. These measures include diversifying Indian agriculture away from production of traditional food grains and traditional crops. Ever since, much emphasis is being placed on diversification of Indian agricultural sector towards production of high-value crops. However, studies on crop diversification in various parts of India considerably differ in establishing any confirmed and uniform relationship between use of water and extent of diversification. At the same time, studies greatly diverge on the role played by farm size in determining extent of diversification (Mukherjee and Chakrabarty 2008; De and Chattopadhyay 2010). Rather, considerable regional variation in extent of diversification has been reported (De 2000, 2003; Ghosh 2010).

14.2 Objectives

In view of the above, an attempt has been made in this chapter to examine (a) whether and under what conditions there has been greater diversification of cropping pattern, and (b) what are the factors at micro-level which influence the farmers' decision regarding diversification. In particular, it has been tried to examine, among other factors including farm size, the role of irrigation in influencing farmers' decision regarding diversification across regions and at the farm level. All these in turn will enable us to identify the conditions under which there has been a greater degree of diversification and identify the farms that show a greater extent of diversification in the light of their size and efficiency.

D. Roy (✉)
AERC, Visva-Bharati, Santiniketan, West Bengal, India
e-mail: debajitroy2007@rediffmail.com

M. Ghosh et al. (eds.), *Diversification of Agriculture in Eastern India*,
India Studies in Business and Economics, DOI 10.1007/978-81-322-1997-2_14

Table 14.1 Basic irrigation characteristics of selected agro-economic regions. (Source: Field Survey)

Agro-economic zones	Districts	Administrative blocks	Irrigational characteristics	Major source of irrigation
Advanced	Barddhaman	Golshi	Highly irrigated	Government canals
Moderately advanced	Birbhum	Bolpur-Sriniketan	Partially irrigated	Government canals and private ground water
Backward	Birbhum	Rajnagar	Nonirrigated	None

14.3 Methodology

Since the aim of this study includes examining extent of diversification in various agro-economic conditions, it covers three purposely chosen agro-economic regions in West Bengal. It should be noted here that the selected regions of the study approximate a highly advanced, a moderately advanced and a backward type of agriculture, varying primarily in the type and degree of availability of irrigation. While one of the regions represents a highly advanced agrarian economy endowed completely with public canal irrigation sources, another in contrast represents backward agrarian economy completely deprived of irrigation from any source. In between, another region represents a moderately advanced agrarian economy, partly endowed by public canals, private tubewells and private river lift irrigation. A brief description of the different regions representing their basic irrigation characteristics is presented in Table 14.1.

The extent of crop diversification at a given point of time may be examined by using several indices. Among these indices, the Herfindahl Index (HI), Simpson's Index (SI) and the Entropy Index (EI) are widely used in the literature of agricultural diversification. All these indices are computed on the basis of proportion of gross cropped area (GCA) under different crops cultivated in a particular geographical area. In this study, Herfindahl Index (HI) has been employed to obtain diversification indices for each of the 450 farms covered under study.

14.4 Results of the Study

14.4.1 Extent of Diversification at the District Level

To work out the diversification indices for the districts at different points of time, we have analysed available aggregative data from official secondary sources for the districts to which the selected agro-economic regions belong. It is observed that both the districts (viz. Burdwan and Birbhum) show increase in their respective diversification indices in 2006/2007 over 2002/2003 with comparable magnitude of change in diversification indices (Table 14.2).

Table 14.2 Surface water irrigation and crop diversification index for the sample districts. (Source: Calculated from District Statistical Handbooks of Burdwan and Birbhum; Bureau of Applied Economics and Statistics, Government of West Bengal 2007)

District	Year	Simpson's crop diversification index (%)	Proportion of irrigated area	
			Canal irrigation (%)	Surface water irrigation (%)
Burdwan	2002/2003	37.36	89.6	93.2
	2006/2007	42.23	89.3	92.8
Birbhum	2002/2003	35.82	65.2	81.6
	2006/2007	40.63	58.3	67.0

Table 14.3 Distribution of sample farms. (Source: Filed Survey)

Diversification category (%)	Avg. value of SID (%)	No. of farms	Percentage of farms (%)
0	0.00	181	40.22
>0–20	10.73	172	38.22
>20–40	28.22	70	15.56
>40–70	50.18	27	6.00
Total	11.50	450	100.00

However, during the same period, it can be observed that proportion of area irrigated by canal has decreased for both the districts, especially registering a sharp fall for district Birbhum, where both of our moderately advanced and backward regions fall. In fact, taking into consideration of all surface water sources (including canal, tank, etc.), it can be observed that the proportion of irrigated area served by surface water sources has declined over the concerned period (especially for district Birbhum), which in turn indicate relative increase in availability of ground water irrigation.

From this very aggregative district level data, we get indications that in reality, crop diversification is not accompanied with relatively less use of ground water. In fact, availability of low cost surface irrigation water does not seem to induce farmers to diversify cropping pattern; rather less availability of such water appears to lead to more use of ground water from private irrigation sources and more crop diversification. These indications will be put under rigorous examination on the basis of detailed data at the level of farming households collected through field survey.

14.4.2 Extent of Diversification at the Farm Level

If we now consider the individual diversification indices, the sample farmers in our study, and stratifying them based on their diversification indices (Table 14.3), we find that proportionate representation of sample farms steadily declines as we move to a higher stratum of diversification index. In fact, while about 40.22 % of sample

Table 14.4 Extent of diversification among farms in different regions. (Source: Filed Survey)

Diversification category (%)	Advanced	Moderate	Backward
0	64 (42.67)	43 (28.67)	74 (49.33)
>0–20	72 (48.00)	49 (32.67)	51 (34.00)
>20–40	13 (8.67)	39 (26.00)	18 (12.00)
>40–70	1 (0.67)	19 (12.67)	7 (4.67)
Total	150 (100.00)	150 (100.00)	150 (100.00)

Figures in parenthesis indicate percentages

Table 14.5 Distribution of diversifying and specializing farms in different regions. (Source: Field Survey)

Type of farms	Agro-economic zone			
	Advanced	Moderate	Backward	All regions
Diversifying	86 (57.33)	107 (71.33)	76 (50.67)	269 (59.78)
Specializing	64 (42.67)	43 (28.67)	74 (49.33)	181 (40.22)
All	150 (100.00)	150 (100.00)	150 (100.00)	450 (100.00)

Figures in parenthesis indicate percentage

farms (181 farms of 450) turn out to be non-diversifier farms, then only 6% of sample farms falls under the highest diversification index category.

However, region wise distribution of the farms according to diversification category shows that majority of the non-diversifying farms come from the backward region (almost unirrigated), followed by the advanced region (completely irrigated with canal irrigation). In particular, about 42.67% of farms belonging to advanced region are found to be non-diversifier farms, as compared to only 28.67% of farms belonging to moderately advanced region.

It is more peculiar to find out that only 1 farm from the advanced region falls under the highest diversification category as compared to 19 farms from the moderately advanced region. In fact, it comes out that an overwhelming majority (more than 70%) of diversifying farms falling under the highest diversification category come from the moderately advanced region.

Clubbing all farms who diversify their cropping pattern as 'diversifying farms' as against the non-diversifying (or 'specializing') farms, we find that majority of the diversifying farms come from the moderately advanced region (Table 14.5), while majority of the specializing farms belong to the backward region. In fact, almost half of the farms belonging to backward region are specializing farms, which results from nonavailability of either surface water or ground water irrigation.

Interestingly, a size-class-wise distribution of diversifying and specializing farms (Table 14.7) reveals that the proportion of diversifying farms increases with increase in farm size. We, however, have found earlier that it is the small farms for whom the average-value of diversification is higher. As such, comparing the two results, it follows that though a higher proportion of larger farms opt for diversification of cropping pattern, but the extent to which they diversify is lower as compared to smaller farms.

Table 14.6 Distribution of diversifying and specializing farms over size-class. (Source: Field Survey)

Diversifier class	Size-class				
	Marginal	Small	Semi-medium	Medium	All
Diversifying	89 (41.78)	83 (73.45)	63 (74.12)	34 (87.18)	269 (59.78)
Specializing	124 (58.22)	30 (26.55)	22 (25.88)	5 (12.82)	181 (40.22)
All	213 (100.00)	113 (100.00)	85 (100.00)	39 (100.00)	450 (100.00)

Figures in parenthesis indicate percentage

Table 14.7 Average-value of diversification index over size-class for diversifying farms. (Source: Filed Survey)

Size-classes	Agro-economic regions			
	Advanced	Moderate	Backward	All regions
Marginal	17.18	29.67	26.76	25.26
Small	12.93	21.64	20.72	18.76
Semi-medium	9.23	24.26	13.35	15.33
Medium	9.21	16.71	10.88	11.91
All	12.80	24.87	18.61	19.24
F-statistic	4.827**	3.327*	4.854**	11.548***

***, ** and * denote significant at 0.001, 0.01 and 0.05 levels

If we now exclude the 181 specializing farms from our analysis (as their presence may dampen the average-value of diversification indices) and consider only the remaining 269 diversifying farms, it still comes out that the average-value of diversification indices for the size-classes declines over increase in farm size. The pattern so revealed comes out to be statistically significant and confirmed by the values of F-statistic in respective analysis of variance exercises.

14.4.3 Factors Influencing Extent of Diversification at the Farm Level

It is obvious from the above results that extent of diversification may vary across farms substantially. The farms bear different characters in terms of size, availability and source of irrigation, as also access to institutional supports like credit. Naturally, the problem is to examine to what extent variation in extent of diversification can be explained in terms of variation in different factors that might influence extent of diversification of farms. Regression analysis is the only way to solve the problem. However, the ordinary least square (OLS) method is inappropriate here because the dependent variable, series of diversification indices expressed in percentage terms for individual farms, is censored, i.e. lies between 0 and 100. Therefore, we have

Table 14.8 Factors influencing diversification for diversifying farms: Tobit regression analysis. (Source: Filed Survey)

Model summary	No. of observations	LR $Chi^2(5)$	Prob>Chi^2	Pseudo R^2
	269	41.86	0.0000	0.0192
Log likelihood = −1069.31				
Independent variables	Coefficients	Std. error	T	P>|t|
Farm size**	−1.496	0.494	−3.03	0.003
Family member per unit of land**	0.452	0.153	2.96	0.003
Dummy for access to formal credit	0.492	1.697	0.29	0.772
Proportion of GCA irrigated by surface-water**	0.008	0.025	0.30	0.765
Proportion of GCA irrigated by ground water	0.077	0.028	2.75	0.006
(Constant***)	17.347	2.456	7.06	0.000
SE	12.887	0.555	–	–

***, ** and * denote significant at 0.001, 0.01 and 0.05 level

considered the Tobit model, which handles truncated and skewed data using the maximum likelihood (ML) method.

The standard Tobit model for truncated and censored variable is defined as:

$$y_i^* = \beta x_i + \varepsilon_i$$

$$y_i = y_i^* \text{ if } y_i^* > 0, \text{ and}$$

$$y_i = 0, \text{ otherwise}$$

where, $\varepsilon_i \sim N(0, \sigma^2)$, and β are the vectors of the explanatory variables and unknown parameters, respectively. The y_i^* is the latent variable and y_i is the dependent variable. The dependent variable y_i is defined here as the Simpson's diversification indices (in percentage terms) of individual farms, and thus what we explain is the factors behind variation in the extent of diversification of farms.

In our present Tobit regression model, we have considered several factors that might influence the extent of diversification among individual diversifying farms. Apart from size of farm measured in hectares, other factors include number of family members per unit of land, and a dummy variable representing access to formal credit institutions (assigning 1 if the farm has accessed formal credit institutions during the year, else 0). At the same time, we have also taken into account the factors like availability and source of irrigation by introducing such variables representing proportion of GCA irrigated by surface water sources and proportion of GCA irrigated by ground water sources. The result of the Tobit regression model thus carried out is present here in Table 14.8.

The Tobit regression exercise performed here turns out to fit to the model specified, as expressed by the values of Chi^2. From the values of the estimated coefficients and corresponding t-values, it comes out that farm size has a significant negative relationship with extent of diversification, while availability of family labour per unit of land shows a significant positive impact on extent of diversification. Interestingly, the impact of availability of ground water irrigation exhibits a significant positive influence on extent of diversification, while availability of surface irrigation does not show any impact. However, the dummy variable introduced to represent access to formal credit institutions does not seem to have any impact on the extent of diversification at the farm level for the diversifying farms.

Hence, the outcome of the present model supports the argument in favour of small farms as diversifier farms, as also supports the fact that availability of family labour plays a positive role in diversification of cropping pattern towards nontraditional crops. At the same time, the results clearly show that it is the availability of ground water sources (and not the availability of cheap surface water sources) that induce farmers to diversify their cropping pattern. It thus appears that it is the smaller farms with abundant family labour who attempt to maximise their return through diversification, even if it requires a higher investment for ground water irrigation. However, the role of availability of formal credit does not show any bearing with the extent of diversification, at least in the present study.

14.5 Conclusion

To sum up, the present study finds that those areas which are endowed with assured water supply at cheap rates (mostly canal irrigation), concentrate more on production of cereals and traditional crops showing lower extent of crop diversification and greater extent of specialization. Again, in those areas where farmers have to depend on private sources of water at high cost due to nonavailability of publicly supplied irrigation system, they diversify away from water-intensive cereals and other traditional crops towards various other high-value crops showing greater degree of diversification. Further, diversification in cropping pattern has occurred more in those cases where farm households are in a position to provide more family labour for cultivation. However, farms' access to institutional credit does not reveal any significant impact on the extent of diversification.

References

De UK (2000) Diversification of crop in West Bengal: a spatio-temporal analysis. Artha Vijnana XLII(2):170–182

De UK (2003) Economics of crop diversification. Akansha Publishing House, New Delhi

De UK, Chattopadhyay M (2010) Crop diversification by poor peasants and role of infrastructure: evidence from West Bengal. J Dev Agric Econ 2(10):340–350

Ghosh BK (2010) Essence of crop diversification: a study of West Bengal agriculture during 1970–1971 to 2004–2005. J Dev Agric Econ 2(11):368–381

Mukherjee A, Chakrabarty S (2008) A study on transformation of Indian agriculture in the post Doha scenario and its macro economic impact. SANEI (South Asia Network of Economic Research Institutes), Global Development Network

Chapter 15
Causes of Agricultural Diversification in Bihar and Jharkhand

Rajiv Kumar Sinha

15.1 Introduction

The finality of diversification, particularly in the wide perspective of agriculture, cannot be confined to 'shift of crop-growing activities' and to 'allied sector based activities or enterprises' only. However, it provides scope for loose fitting of shifts or change in areas within crops, or particular types of activities also. Diversification is a widely advocated means for agricultural and rural development. In agriculture, diversification implies shift from subsistence farming to commercial farming and/or from low value to high value food or non-food crops and switch over from local to high yielding varieties. It also includes shift from agriculture to animal husbandry, pisciculture, poultry, agro-forestry, horticulture, etc. or vertical diversification, i.e. shift to non-farm economic activities, or shift to such 'agro-product based commercial activities' that enable the farmers to get higher net returns with lower degree of risks involved. Change in cropping pattern is an integral part, and a popular mode of diversification, which has far-reaching impacts on development of agriculture and alleviation of rural poverty (Ram 1999).

During the past century, an unprecedented increase in population has been experienced by the world with a billion people added every decade during the last more than three decades (1978–2011). Dramatic shifts in production and consumption of food have accompanied this population explosion including a surge in grain production, spectacular rise in meat production and consumption, and the emergence of an increasingly vital role for international trade. Despite several measures to check population explosion, India's population is still rapidly increasing. The standards of living (both in urban and rural areas), have been rising and slated to rise faster than in the past. It can be practically dilated that as the living standards rise, both calorific intake and diversification of diet will, no doubt, increase

R. K. Sinha (✉)
Agro-Economic Research Centre for Bihar and Jharkhand, Old P.G. Campus,
T.M. Bhagalpur University, Bhagalpur 812007, Bihar, India
e-mail: rajiv.sinha1959@gmail.com

M. Ghosh et al. (eds.), *Diversification of Agriculture in Eastern India*,
India Studies in Business and Economics, DOI 10.1007/978-81-322-1997-2_15

significantly. Although a portion of this increase is likely to be obtained from abroad, a fulfilling 'Vision of India, 2020' evidences India with its conducive and varied climate, the largest irrigated area in the world, and a vast farming population as a major food exporter.

In the endeavour to achieve the goal of food security for all our people and with the objective to earn foreign currency by exporting agro-based products, contriver measures may be adopted amidst varied climate and increasingly diversified diet. Both as a challenge and as an opportunity, India can and must do much more to modernize and diversify its agriculture to meet the increased domestic and international demand for a wide variety of food products. Though agricultural diversification in India has been all about a continuous increase in the share of allied activities and decline in the share of crop sector since the late 1960s, however, within crop sector, trend of diversifications has been changing periodically. In spite of these periodic shifts in crop acreage, proportionate area under fine cereals has been increasing over a long period. There can be various reasons for such trend. Though in some of the states, this trend/practice continues, the recent trend in crop diversification shows decline in percent area under fine cereals and increase in percent area under non-food grain crops (Kalamkar 2008).

During the 8 years' period (from 2000–2001 to 2008–2009), percentage changes of coverage under different crops in Bihar suggest diversification of pulses, fibre crops, oilseeds and cereals towards sugarcane (19.64 % increase in area). In case of Jharkhand, diversification of agriculture within crops during the 8 years' period (2001–2002 to 2008–2009) disillusions farmers' preference towards paddy (with only 9.85 % increase in area), and suggests the farmers to grow fascination towards oilseeds, pulses, maize and wheat (74.12, 73.66, 54.46 and 52.75 %), respectively. Diversification of agriculture within crops in Bihar and Jharkhand might have been caused due to some constraints such as natural, agro-ecological factors, absence of remunerative and exploitation-free markets for surplus production, lack of adequate storage facilities, non-availability of sufficient number of agro-processing industries, etc.

In the above backdrop, and based on available secondary data, this paper seeks to examine the causes of agricultural diversification in Bihar and Jharkhand. The specific objectives of this paper are to (i) examine the diversification within crops in Bihar and Jharkhand, (ii) find out factors responsible for such crop diversification, (iii) discuss the effects of such diversification and (iv) suggest appropriate actions.

15.2 Factors Responsible for Diversification

There are several factors responsible for diversification of crops, diversification in regard to area; activities and otherwise. It can vary across the regions, states and 'agro-climatic zone wise'. During the last 25 years and more, the diversification of agriculture in India has remained a big issue. Much progress has not been made in this regard due to various technological, social, economic and institutional factors.

The eastern region of the country has about 77 % of the gross cropped area (GCA) under food grains. No doubt, food grains production more than doubled from 102 million t in triennium ending 1973 to 227.3 million t (MTs) in 2007–2008. Virtually all the increase in production resulted from yield gains, rather than expansion of cultivated area under food grains that remained stagnant at around 124 million ha (MH) in India (Sidhu 2006). However, per capita per day availability of food grains in India could also be seen to be stagnant. It was 468.8 gm/capita/day in 1971 that decreased to 462.4 in 2004. The strength of India being one of the leading nations of the world in regard to production of fruits, vegetables and milk could have been utilized to reduce the extent of poverty by creating additional employment opportunities based on 'agro-processing activities/industries' of these products (Jha 2007).

Despite some of the impressive growth and development on quite a few fronts in recent past, India is still home to the largest number of poor people in the world. With more than 251 million below poverty line population, India accounts for about one-fifth of the world's poor population(Kumar 2005). Because of high population pressure, India's land and other resources to meet its food and development needs are under severe pressure. These all factors (combined together in various proportions) have led to change in the structure of food basket. No doubt, diets are diversifying from basic cereals to fruits, milk and milk products, meat, fish and eggs, etc. Similarly, the population growth, rise in per capita income, urbanization, change in taste and preferences, overall economic growth and infrastructural development could be responsible for change in the supply and demand for food items.

Now, it is a matter to be seen and discussed that apart from the above factors directly or indirectly responsible for 'diversification of agriculture' on all India level, what are the constraints/factors that could be instrumental for 'diversification of agriculture within crops' in the states of Bihar and Jharkhand.

15.3 Diversification Within Crops in Bihar and Jharkhand

Data pertaining to area under cereals, pulses, oil seeds, fibre crops and sugarcane for the period from 2000–2001 to 2008–2009, percentage change in area under these crops in the year 2008–2009 over 2000–2001, and percentage change in triennium average (2006–2007 to 2008–2009) over (2000–2001 to 2002–2003) have been taken into consideration for assessing the increase/decrease in area under these crops, and to see the trend towards specialization or diversification.

A glance on Table 15.1 reveals that the area under cereals (taken as a whole), declined by 66,410 ha during the period 2000–2001 to 2008–2009. In Bihar, it was 6400.15 thousand ha and it came down to 6333.74 thousand ha in 2000–2001. It means there was a decline of 1.04 % in area under cereals in 2008–2009. As far area under pulses in concerned, there was a significant decline of 130,510 ha during this period. It came down from 714.88 thousand ha of 2000–2001 to 584.37 thousand ha in 2008–2009. Thus, an 18.26 % decline in area under pulses could be seen in the

Table 15.1 Area under major crops in Bihar (area in '000 ha). (Source: Economic Survey (Bihar): 2010–2011, Government of Bihar)

Year	Cereals	Pulses	Oilseeds	Fibre crops	Sugarcane	Total
2000–2001	6400.15	714.88	153.70	169.66	93.53	7531.92
2001–2002	6362.26	694.26	147.89	160.56	113.44	7442.41
2002–2003	6346.45	696.88	137.23	172.07	107.27	7459.90
2003–2004	6282.71	680.88	140.53	178.04	103.60	7385.76
2004–2005	5838.80	658.06	131.88	154.39	101.24	6884.37
2005–2006	5959.71	566.94	137.90	148.77	104.19	6917.51
2006–2007	6237.12	610.07	143.11	154.30	117.18	7261.78
2007–2008	6304.68	581.50	142.05	154.25	107.04	7289.52
2008–2009	6333.74	584.37	138.08	150.97	111.90	7319.06
% change in 2008–2009 over 2000–2001	−1.04	−18.26	−10.16	−11.02	19.64	−2.83
TA (2000–2001 to 2002–2003)	6357.62	702.01	146.27	167.43	104.75	7478.08
TA (2003–2004 to 2005–2006)	6027.07	635.29	136.77	160.40	103.01	7062.55
TA (2006–2007 to 2008–2009)	6291.85	591.98	141.08	153.17	112.04	7290.12
% change in TA (2006–2007 to 2008–2009) over TA (2000–2001 to 2002–2003)	−1.04	−15.67	−3.55	−8.52	6.96	−2.51

Percentage change, triennium average (TA) and percentage change in TA (2006–2007 to 2008–2009) and over TA (2000–2001 to 2002–2003) have been calculated on the basis of data presented in above table

state. Oilseed crops witnessed a decline of 15,620 ha only during the period under review. It fell down from 153.70 thousand ha in the initial year to 138.08 thousand ha in the year 2008–2009. In physical terms, it might not be much larger, but in percentage terms, the decline in 'oilseeds area' was as high as 10.16. In case of fibre crops, a little more decline than the oilseed crops could be seen. There was a decline of 18,690 ha in area. It was 169.66 thousand ha. in 2000–2001 that came down to 150.97 thousand ha in 2008–2009. In percentage terms, the decline was of 11.02 (Economic Survey, Bihar 2010–2011).

Contrary to above, the area under cash crop (sugarcane) increased significantly in the state during that period and there was an increase of 18,370 ha in its area, which was 93.53 thousand ha in 2000–2001. This area increased to 111.90 thousand ha in 2008–2009. In this way, an encouraging increase of 19.64 % could be noticed in the area under sugarcane. Having analysed data in the Table 15.1 (en-masse), analysis may lead to divulge that nearly 7.95 % of the total area is under cereals, pulses, oilseeds and fibre crops have shifted towards sugarcane. However, on overall level, the percentage decline in areas of all crops (taken together) was to the lower extent of 2.83 during the period. But, it was 2,12,860 ha in area. Actually, area under all crops, that was 7531.92 thousand ha in 2000–2001, declined to 7319.06 thousand ha in 2008–2009 (Table 15.1).

Table 15.2 Area under major crops in Jharkhand (area in '000 ha). (Source: (i) Crop-wise data for the year 2001–2002 to 2008–2009 have been obtained from the Directorate of Agriculture, Government of Jharkhand, Ranchi and (ii) percentage change, figures on TA and percentage change in TA have been calculated on the basis of data presented in above table)

Year	Paddy	Wheat	Maize	Pulses	Oilseeds	Total
2001–2002	1520.61	65.38	139.88	211.64	74.96	2012.48
2002–2003	1383.23	67.88	157.60	242.95	94.41	1946.08
2003–2004	1363.86	74.56	187.51	301.89	101.07	2028.91
2004–2005	1276.42	64.50	191.24	290.91	94.26	1917.34
2005–2006	1354.72	57.90	181.24	291.07	84.84	1969.88
2006–2007	1623.62	84.31	240.86	376.63	144.68	2470.11
2007–2008	1643.78	86.34	237.41	393.66	124.34	2485.54
2008–2009	1670.33	99.84	216.06	367.53	130.52	2484.33
%Change in 2008–2009 over 2001–2002	9.85	52.75	54.46	73.66	74.12	23.45
TA (2003–2004 to 2005–2006)	1331.67	65.68	186.66	294.62	93.39	1972.04
TA (2006–2007 to 2008–2009)	1645.91	90.17	231.44	379.27	133.18	2480.00
% Change in TA (2006–2007 to 2008–2009) over TA (2003–2004 to 2005–2006)	23.60	37.29	23.99	28.73	42.61	25.76

TA triennium average

Apart from other factors, some of the identified state specific constraints that could have possibly led to such decline/diversifications within cropping pattern in Bihar can be mentioned. Low productivities of major crops of rice and wheat are due to low seed replacement ratio, flash floods causing inundation; and pest and diseases attacks. It is to be noted here that due to lack of assured irrigation, storage/godown facilities and un-remunerative prices of cereals (in case of good harvests), etc., particularly in case of maize, and in absence of desired number of APIs in the state, the areas under cereals and pulses have shifted towards low cost crops.

Further, water resources sector is most vital for the development of Bihar. It would determine the trend of diversification in Bihar. No doubt, the problem of development and management of water resources is highly complex. It is more so in case of Bihar as the state encounters different kinds of problems, sometimes opposite in nature. Flood, drought and water logging occur frequently in 73, 17 and 10% of the geographical areas of the state respectively. Flood is a big menace, mainly in North Bihar, and agriculture is badly affected, though the land is very fertile. Frequent flood in major part of north Bihar has also led to diversification of quite a large area towards 'non cereals crops', particularly during kharif season (Anonymous 2009).

Having a look on data in Table 15.2, it reveals that in Jharkhand, there was a marked increase in areas under paddy, wheat, maize, pulses and oilseeds during the 8 years' period from 2001–2002 to 2008–2009. Percentage changes in acreage in 2008–2009 over 2001–2002 for these crops could be estimated at 9.85, 52.75, 54.46,

73.66 and 74.12 % respectively. On overall level, the percentage change in area under all crops (taken together) is calculated at 23.45 %. Paddy, the main cereal crop of Jharkhand had witnessed increase of 1,49,720 ha during this period. Out of the total geographical area of 79,70,080 ha, 23,33,550 ha (29.28 %) is under forest area. As per the statistical profile of Jharkhand (2006), land put to non-agricultural practices, barren and unutilized land, permanent pasture and other grazing land, cultivable wasteland, land under miscellaneous trees, current fallow, net area sown, area sown more than once were estimated at 6,88,270 ha (8.64 %), 5,75,780 ha (7.22 %), 87,460 ha (1.10 %), 2,83,620 ha (3.56 %), 1,24,270 ha (1.56 %), 13,63,050 ha (17.10 %), 17,62,470 ha (22.11 %) and 2,63,040 ha (3.30 %), respectively.

In Jharkhand, crops are grown in four seasons, namely Bhadai, Aghani, Rabbi and Garma. However, total irrigated areas under different crops in each of the four seasons were 1971.60, 68,357.36, 74,827.36 and 12,302.33 ha, respectively. Total irrigated area under different crops in the four seasons was 1,57,458.65 ha. Only 8.93 % of the net sown area is having irrigation facility in Jharkhand. In 'such irrigation scarce situation', growing wheat, paddy, rabi maize and garma paddy like cereal crops is not an easy to operate business; rather it may be an sceptic activity. But the other side of the scenario is also to be kept in mind that Jharkhand is a state with undulated topology, hill-hillocks, mountain ranges and hilly rivers with almost no water during the whole year except rainy season (for shorter duration only; Jharkhand: A Statistical Profile, 2006).

The above factors provide sufficient ground to understand that in a large part of the state, particularly in the foothills and upland areas around long and small mountain ranges, paddy, arhar, moong, ghaghra like pulse crops can only be grown as rainfed crops. A substantial fact here to be urgently noted is that with rapid deforestation in the state, areas under paddy, maize, pulses and oilseeds might have increased to some extent.

Besides the above constraints and specific natural characteristics, Jharkhand is also cursed with the following state specific constraints. These are lack of irrigation facilities, large 'rice fallow area' (75 % of NSA)', which remains uncultivable in rabi season due to lack of irrigation facilities, low remunerative upland rice production, acidic soils, and low weed replacement ratio. Despite these constraints, there are also evidences of diversification within crops. It means larger increases in areas under oilseeds (74.12 %), pulses (3.66 %), maize (54.46 %), wheat (52.75 %) and comparatively lower increase in area under paddy (9.85 %). These constraints can be minimised by: (i) construction of irrigation and 'rain water harvesting structures' in the forms of (a) tube well irrigation and (b) dug well to increase the irrigated area and ensure water availability with the help of convergence with micro irrigation, (ii) intensive cultivation of rice for 17 Non-NFSM (National Food Security Mission) districts of Jharkhand, (iii) intensive pulses production in NFSM districts, (iv) maize and wheat development programme and (v) bridging knowledge gaps/training. These endeavours of the government through its programmes/specific schemes could have led to 'diversification of agriculture within crops' in terms of greater or lower increases in areas of different crops in Jharkhand.

15.4 Suggested Action Points

In view of the likely rise in calorific intake and 'diversification of diet (as a result of rise in general income level' and various disease-related compulsions policy makers, agricultural research centres and the Government will have to let the farmers know/advise to undertake shifts in cropping pattern (as per the demand of the domestic and international markets). With the objectives of generating additional income by enhancing employment opportunities in rural areas (on sustainable basis), farmers need to be incentivised and encouraged to undertake those non-food crops also, which have greater potentiality of value-addition. In view of the fact that major part of the increase in production had resulted from yield gains rather than expansion in area, special efforts need to be made for increasing productivities of major crops in eastern region of India by promoting use of green manures and cow dung.

It is more so desirable with a view to save cultivable land from degradation to save it from the threat of being barren and sustain fertility of top soil. Nearly 8 % shift of areas under cereals, pulses, oilseeds and fibre crops (taken together) is towards sugarcane. So, processing industries (based on particularly this crop) need to be rejuvenated and installed in a good number in Bihar in the areas/regions of its surplus production. By adopting both structural and non-structural measures for the management of flood, and through development of Bihar's mega water resources assured irrigation coverage can be sufficiently increased. Watershed development programmes/initiatives will have to be largely carried out in foothills and uplands of mountain ranges of Jharkhand. It is desirable to expand 'ensured irrigation facility' in 'large rice fallow area' in this state (nearly 75 % of NSA). Constructed rainwater harvesting structures need to be surrounded by plants and trees (of fruits), so that incidence of 'loss of water through evaporation' may be reduced and 'deforested areas' in Jharkhand could be compensated to some extent. Besides irrigation purpose, such 'water reservoirs' should be sued for fisheries and duckery also. In view of very poor irrigation facility in Jharkhand, 'allied agricultural activities', (viz. livestock, dairy, piggery, poultry, fishery, bee keeping, silkworm rearing, APIs based on Minor Forest Produces, i.e. MFPs should be emphasized, promoted and developed. If as a result of diversification within crops, such excess production of a particular crop is achieved, which has no local, domestic and/international market demand at 'at least remunerative prices', and which has no 'value-addition facility, at the local/regional level, in that case, it will be advisable to encourage the farmers to undertake allied agricultural activities in those particular areas of their land. If low cost crops are getting good market demand or are able to bring enhancement in 'the general standard of living of the farmers', then such diversification within crops' may be allowed.

References

Anonymous (2009) Resurgent Bihar—a development vision (Report of the special task force on Bihar). Government of India, New Delhi, Jan

Economic Survey (2010–2011). Government of Bihar

Government of Jharkhand (2006) A statistical profile. Directorate of Statistics & Evaluation, Jharkhand, Ranchi

Jha B (2007) Was crop diversification responsible for the recent food crises in India? Agricultural Situation in India (Special Number), Aug, pp 341–349

Kalamkar SS (2008) Food security in India in global perspective. Agricultural Situation in India (Special Number), August, pp 291–302

Kumar P (2005) Empowering the small farmers towards a food secure India, In: India's agricultural challenges: reflections on policy (eds). Technology and other issues. Centad Publications, New Delhi, pp 197–225

Ram S (1999) Cropping pattern diversification in Orissa. Agric Situat India LVI(1):15

Sidhu MS (2006) Diversification of agriculture in East India: some issues. Paper presented in a national seminar at Patna (26–28th March, 2006) jointly organized by IEA & EAB, pp 1–9

Chapter 16
Policy Intervention in West Bengal Agriculture: Role of Diversification

Shiv Raj Singh, Subhasis Mandal, K. K. Datta and Uttam Bhattacharya

16.1 Introduction

The eastern region of India, with its diverse natural resources, has been under special focus by the Planning Commission for ushering the second green revolution in the country. The eastern states, Bihar, Orissa, Jharkhand, and West Bengal have plenty of natural resources as well as diverse agro-eco-regions for increasing the food production of the country. Besides producing food grains, these regions have great potential to produce several high-value commodities for augmenting farm income. For example, horticulture, livestock, and fisheries sectors have all potential to pull the growth of agriculture sector of this region. One of the key impediments of fostering the agricultural growth in these regions is the small and marginal production unit of the majority of the farmers. The small scale of the production unit produce these high-value commodities with high to moderate production efficiency, but severely constrained with poor marketing efficiency. Farming units are confronted with many unpredictable uncertainties ranging from climatic vagaries to market price fluctuation. The degree of uncertainty is greater to the small and marginal landholders where the farmers do not have access to basic information on various

S. R. Singh (✉)
Shri G.N. Patel Dairy Science & Food Technology College, SDAU, S.K. Nagar, Gujarat
e-mail: shivagritech2007@gmail.com

K. K. Datta
National Dairy Research Institute (Deemed University), Karnal 32001, Haryana, India
e-mail: kkdatta2007@gmail.com

S. Mandal
Central Soil Salinity Research Institute, Canning Station, West Bengal, India
e-mail: subhasis2006@gmail.com

U. Bhattacharya
Institute of Development Studies Kolkata (IDSK), 27/D Block DD, Sector-1, Saltlake, Kolkata 700064, West Bengal, India
e-mail: uttam.bhattacharya@gmail.com

M. Ghosh et al. (eds.), *Diversification of Agriculture in Eastern India*,
India Studies in Business and Economics, DOI 10.1007/978-81-322-1997-2_16

risks including loss of assets and income. The small and marginal farmers are more vulnerable to risks arising from the very limited resources and assets at their disposable. As uncertainly is an important consideration in household decision-making, it enters the utility function of the farm household as a preference for security in the face of several risks including the loss of their investment, low income, and food shortage. In general, utility maximization involves a trade-off between higher income and greater security; whereas, in case of small and marginal landholders these two are separate goals. Keeping in view the opportunities, sources of growth, and prospects, the main purpose of this chapter is to focus on the ways and means of agricultural development in West Bengal, which may help to accelerate the rural income, household level food, and nutritional security.

16.2 Structural Changes in West Bengal Agriculture

The progress of West Bengal agriculture during the last decade has been slow and volatile, which has raised concerns in terms of food and nutrition security. At the national level the overall annual agricultural growth during the Xth Five Year Plan (2002–2007) was about 2.5 % against a target of 4 %, and the first 3 years of the XIth Five Year Plan also clocked a similar growth of 3.1 %. In case of West Bengal, the growth rate in agriculture and allied sector has been recorded as 2.31 % and within the sector the agriculture has recorded a slow growth of just over 2.01 %, whereas the overall growth rate has been 6.06 % during the period from 1999–2000 to 2007–2008 (Table 16.1). The percentage share of agriculture is gradually declining, which is a natural phenomenon under economic transition of any state/country. On the other hand, in the livestock sector, against a targeted growth of 6 %, the realized growth was 4.17 % per annum. Over the planned period, the annual growth rates for livestock and horticulture sectors were faster relative to the growth rates of the agriculture sector. This growth had a smoothing effect in the total growth rates of the agriculture sector. In order to accelerate growth in agriculture and to bring about stability in agricultural growth, to ensure food, and nutrition security, there is a need for some policy and institutional changes.

Indian economy has recorded a spectacular growth of over 8 % during the post-liberalization period since 1991. However, the performance of West Bengal's economy has been relative poor as the state has failed to take advantage of the ongoing economic liberalization process. Similarly, the per capita income (PCI) of West Bengal has gradually slipped down as compared to other states, and currently, the income level is (₹ 31722) even below the national average (₹ 33283) at current prices. The PCI growth was 2.57 % during 1980–1981 to 1993–1994, which accelerated to 5.56 % from the period 1993–1994 to 1999–2000 , but slowed down to 4.80 % from 2000–2001 to 2007–2008. In terms of all India rank of net state domestic product of West Bengal, it more or less gained third place, but the all India rank in relation to PCI is steadily slipping down from 7th in 1980–1981

Table 16.1 Percent share of various components of NSDP at constant prices (Base year 1999–2000). (Source: CSO, Government of India)

Components	1999–2000	2003–2004	2004–2005	2007–2008	ACGR
Agriculture	27.4	24.5	23.2	20.3	2.01
Forestry and logging	0.8	0.9	0.9	0.9	7.82
Fishing	3.5	2.9	3.0	2.8	3.12
Agriculture, forestry and fishing total	31.7	28.4	27.0	24.1	2.31
Industry	10.6	11.0	11.3	9.9	5.40
Services	57.7	60.7	61.7	66.0	7.90
Total NSDP (crore)	125,299	152,384	162,491	201,298	6.06

ACGR: Annual compound growth rate (%)

to 18th in 2006–2007. However, the real concern is the simultaneous slowdown of growth rate of industrial sector. The industrial growth has remained less performing as compared to many other agriculturally important states of India. For example, Andhra Pradesh has recorded 5 % growth in agriculture, 7.2 % in industry, and 8.3 % in service sector during the same period. Since few decades, the agricultural sector is becoming less-performing due to lack of significant technological breakthrough after the green revolution that sometimes has been termed as "technology fatigued" stage, and indeed is a national concern for India. Under this situation, the industrial sector which has complementary relationship with both agriculture and service sector, must perform better to pull the overall growth of West Bengal's economy.

Another feature of the West Bengal's agricultural economy is that it is mainly dominated by the marginal farmers' category—over 80 %—but owns only 50 % of the total area (Table 16.2). The land holdings are still getting fragmented over the years. The average size of holding is only 0.82 ha which is much less than the national average (1.33 ha). Marginal holdings of operational lands often become a constraint to adoption of new technologies and mechanization. Very often, the technologies are not divisible like land fragmentations. Under these circumstances, larger investment becomes uneconomical. Due to the small scale of production the farmers usually produce the crops with high to moderate production efficiency but suffer from poor marketing efficiency due to small quantity of marketable surplus. Many have opined that the high growth in agriculture in West Bengal during 1980s was both by the technological and institutional factors, such as the delayed green revolution, personal initiative, private investment, operation barga, distribution of surplus land, and Panchayati Raj (Sengupta and Gazdar 1998; Banerjee et al. 2002). Some others have argued that the increased land productivity in West Bengal during the 1980s is the direct result of the increased labor intensity in family farms belonging to marginal and small category of farmers. As the land reforms created higher purchasing power, marginal and small farmers could access green revolution technology unlike in the leading green revolution states of India (Bhattacharya and Bhattacharya 2007).

Table 16.2 Percentage distribution of ownership of holdings and owned area in West Bengal (%). (Sources: Agricultural Statistics at a glance—various issues, Ministry of Agriculture, Government of India)

Size class	1972		1982		1992		2002	
	Holdings	Owned area	Holdings	Owned area	Holdings	Owned area	Holdings	Owned area
Marginal	77.62	27.08	81.6	30.33	85.88	41.29	80.44	49.74
Small	12.64	25.69	11.50	28.77	9.48	28.11	14.86	28.97
Semi-medium	7.30	27.72	5.54	27.23	3.94	22.98	4.17	14.13
Medium	2.39	18.61	1.28	12.12	0.71	7.62	0.52	3.21
Large	0.05	0.70	0.08	1.54	Neg.	Neg.	Neg.	3.95

Average holding size in West Bengal is 0.82 ha during 2001 and the same is 1.33 ha for India
Landless 0–0.002 ha, *Marginal:* 0.002–1.00 ha, *Small* 1.001–2.00 ha, *Medium* 2.001–4.00 ha, *Large* ≥ 4.00 ha

Table 16.3 Trends in poverty in West Bengal and India. (Source: Himansu 2007)

State/India	Rural/Urban	1983	1993–1994	2004–2005
West Bengal	Rural	61.56	37.35	28.49
	Urban	31.5	23.24	18.5
All India	Rural	45.76	37.26	29.18
	Urban	42.27	32.56	26.02
West Bengal	Rural + Urban	53.60	33.45	25.67
All India	Urban + Urban	44.93	36.02	28.27

The other interesting feature was that West Bengal has performed reasonably well in reduction of rural poverty as compared to rest of India. During 1983, around 62 % of rural population in Bengal was under poverty and that came down to 28 % during 2004–2005 (Table 16.3); whereas, during the same period the rural poverty at national level declined to 29 % from 46 %. Overall, the poverty level in West Bengal declined to 26 % in 2004–2005 from 54 % in 1983, and the national level the decline was from 45 to 28 % during same period. However, the reduction in poverty in West Bengal was much higher (over 24 percentile points) from the period 1983 to 1993–1994 as compared to the period from 1993–1994 to 2004–2005 (9 percentile points). The better performance during early period was mainly due to the high growth in agriculture and later the agricultural performance slowed down and simultaneously the performance of the industry was also poor, thereby it affected the overall performance in poverty reduction of the state.

16.3 Labor Surplus Economy and High Value Agricultural Subsectors

In the labor surplus economy, it is advisable to generate ample employment opportunities in the rural areas and there is a need to move the rural masses towards high-value agriculture like livestock rearing, fishery, and horticulture. These three subsectors of agriculture system are characterized by the labor-intensive operations (activities) and they are not easily substituted by the mechanization. As we know that in West Bengal mostly workers are engaged in the informal sector where wages and social security is not handsome and secure. Therefore, there is need to engage the huge labor force in the rural setting by using them in high-value agriculture namely livestock rearing, fishery, and horticulture. Smallholder farming system is predominately characterized by the home based labor. In this smallholder farming system, if they move towards the high-value agriculture subsector then they can easily enchase higher wage share for household surplus labor force. As we know, to move towards the commercialization of agriculture there is a need to diversify the agriculture and try to explore its needs, potentiality, and income regenerating ability of the rural masses from its subsistence sources of assets. In case of West Bengal, diversification of agriculture means to integrate the livestock rearing, fishery, and horticultural activities to secure the rural livelihood.

16.4 Livestock Sector

Based on a review of a large number of papers from South Asia, Devendra et al. (2000) reported a paucity of research that incorporates livestock interactively with cropping, and a woeful neglect of social, economic, and policy issues. Biophysical commodity-based crop or livestock research dominated system perspective was lacking, and many of the technologies which were developed were not adopted. More recently, broad classifications of crop–livestock systems in South Asia and their component technologies have been documented (Paris 2002; Thomas et al. 2002; Parthasarathy Rao and Hall 2003). However, it is clear that a better understanding of farming systems and of the livelihood objectives of landed and landless families, including how they exploit crop–livestock interactions, will be required if we are to be successful in improving rural livelihoods and securing environmental sustainability. Labor intensive crops, dairying, and off-farm diversification all contribute to a relatively broad-based growth. Livestocks and vegetable cultivation provide significant complementary income sources with an ample market to tap. The livestock and fisheries sector is growing very fast in India. A recent data shows that during 2008–2009, in India, the livestock and fisheries sector contributed over 4.07 % of the total GDP and about 26.84 % value of the output from total agriculture and allied activities.

Owing to its diverse physical and climatic conditions, West Bengal has a varied livestock population in India. Indigenous livestock resources of West Bengal like Black Bengal goat, Garole sheep, and Ghoongro pig have considerable traits of adaptability to withstand environmental stress. According to the Livestock Census of 2007, West Bengal has about 19.2 million cattle (9.7 % of national population), 0.8 million buffaloes (0.7 %), 15.1 million goat (11.5 %), 1.6 million sheep (2.3 %), 0.8 million pig (7.3 %), and 64 million poultry (21.9 %).

West Bengal alone has about 20 million cattle and buffaloes producing around 4.4 million t of milk, cows contributing the major share (4.1 million t). Majority of the large ruminants belong to indigenous breeds (71 %), which contribute 61 % of total milk production of the state. It may be highlighted that this state owes 9.7 % of national cattle population, which contributes only 3.9 % of total milk production in India. Therefore, it is important to visualize emerging scenario and devise appropriate policies and strategies to ensure the required interventions for development of large ruminants stressing on indigenous breeds.

The national availability of the milk is now about 258 g/person/day which was only 112 g/person/day in 1968–1969, and is getting closer to the world's average of 265 g/person/day. But in eastern India, present availability of milk is around 125 g/person/day, which is about half that of the national average. The pattern of use of milk and milk products in eastern region is almost similar to the national scenario. A total of 52 % liquid milk is used for consumption purpose in West Bengal compared to 48 % milk for production of milk products, mainly for industrial use. As the level of milk production in eastern India is considerably low instead of high demand of milk products (mainly sweets), majority of the milk is procured for industrial use leaving low availability for per capita consumption. Therefore, majority of the milk for consumption purposes are produced through reconstitution of imported milk powder.

16.5 Fisheries Sector

India, with diversified agro-climatic regions, is endowed with potentially rich and varied aquatic resources. It is the third largest fish producing country in the world and ranks second in inland fish production. This sector is often termed as the sunrise sector in Indian agriculture as it has huge potentials to provide nutritional security, rural and livelihood development, employment generation, and export earnings through diversified farming practices, both in the marine and inland sector. Fishing, aquaculture, and allied activities provide livelihood to over 1.4 million people. Further, untapped potentials exist in inland systems from ornamental fishes to value added products. These resources offer immense scope and potential for development of aquaculture and fisheries in India.

Average per capita consumption of fish/fish products in India (11.0 kg/annum) is still low as compared with other developed countries (16.0 kg). The percentage share of fisheries in agricultural GDP was 5.34 % as of year 2004–2005. The average growth rate of inland fish production in India is computed to be 4.29 %. Fish being one of the most favorite food items of people of West Bengal, fishery activities have been carried out in West Bengal from ancient time. A rich diversity of species and favorable climate, cheap labor, and easy distribution make India and West Bengal in particular, suitable for fish culture. But fish culture is practiced in less than 30 % of the total available area. There is a good potential to increase fish production as the demand is increasing rapidly. This can be possible by covering the areas suitable for fish culture through the scientific way of fish farming.

The main inland fishery resources of India include about 1.20 million hectares (Mha) of brackish water area; 2.38 Mha of fresh water ponds and tanks; about

Table 16.4 Production of fruits and vegetables in West Bengal. (Source: Calculated from National Horticulture Database 2008, Ministry of Agriculture, Government of India)

Year	West Bengal			% Share to India		All India yield
	Area (000 ha)	Production (000 t)	Yield (kg/ha)	Area	Production	(t/ha)
1991–1992	111.3	1131.7	10.2	3.87	3.95	10.0
2005–2006	172.7	2301.7	13.3	3.24	4.16	10.4
2007–2008	194.3	2766.7	14.2	3.49	4.36	11.0
1991–1992	456	4680	10.3	8.15	7.99	10.5
2005–2006	1210.3	18,663	15.4	16.78	16.75	15.4
2007–2008	1313.2	22,456.8	17.1	16.83	17.84	16.8

1.24 Mha of reservoirs; 0.82 Mha of beels, oxbow lakes, and derelict water bodies; 0.24 Mha of floodplain wetlands; 0.29 Mha of estuaries; 1.65 Mha of mangroves, swamps, lagoons, etc. besides about 191,000 km of rivers and canals. As far as West Bengal is concerned, it has 2526 km (1.3 %) of rivers, 0.17 lakh ha (0.6 %) of reservoirs, 2.76 lakh ha (11.43 %) of tanks and ponds, 0.42 lakh ha (5.26 %) of flood plain and derelict water bodies, 2.10 lakh ha (17 %) of brackish water area of India. In total, West Bengal has 5.45 lakh ha (7.41 %) of water area for fish production out 73.59 lakh ha of total suitable available water bodies in India. Fish production from inland sector of India rose to 4.7 million metric tonnes (MMT) with contribution from West Bengal at 1.44 MMT (20.30 %). It is good to see that the state of West Bengal with an annual production of about 1.5 million t of fish has emerged as the highest fish producing state in the country. Further, West Bengal contributes to almost 65 % of the seed production of the country. In spite of the vast resources, average production of fish is not at par when compared to other states like Punjab and Andhra Pradesh. There are many factors that contribute to low production of fish apart from demand for "quality," "disease free," fry/fingerlings, which are continuously increasing and needs to be addressed quickly.

16.6 Horticulture Sector

West Bengal is already a leading state in terms of producing vegetables in India. The state is accounted for 17.8 % of total vegetable production from 16.8 % area of India in 2007–2008 (Table 16.4). Fruit crops are also grown widely across the state and contribute 4.4 % of India's fruits production from 3.5 % of total area. The vegetable production in the state has increased sharply from 4.7 million t in 1991–1992 to 18.7 million t in 2005–2006, and further increased to 22.46 million t during 2007–2008. However, the sector is facing steep challenges towards enhancing crop productivity, generating gainful employment, improving economic conditions of the farmers, and building entrepreneurship. In addition, marketing function of high value agricultural commodities such as fruits and vegetables has been very poor and primitive in the state and needs innovation to increase the marketing efficiency

Table 16.5 Household level distribution pattern of dairy animals in West Bengal. (Source: Authors' estimates from the unit level data on Situation Assessment Survey of Farmers, National Sample Survey Organisation (2005) 59th round, Ministry of Statistics and Programme Implementation, New Delhi)

Particulars	Land less	Marginal	Small	Medium	Large	All
Estimated Sample households keeping dairy animals (no. in lakh)	5.69	32.39	2.97	0.81	0.09	41.95
Total Estimated Sample household (no. in lakh)	11.27	53.01	3.84	1.02	0.11	69.25
Percentage of household keeping dairy animal	50.49	61.10	77.34	79.41	81.82	60.58
Number of dairy animals hold by households (%)	11.97	74.48	9.52	3	1.02	100
Percentage of land hold by households	0.01	66.25	20.81	10.39	2.53	100
Average herd size (no.)	1.27	1.68	2.97	3.52	10.67	1.73

Landless 0–0.002 ha, *Marginal*: 0.002–1.00 ha, *Small* 1.001–2.00 ha, *Medium* 2.001–4.00 ha, *Large* ≥ 4.00 ha

and to reduce the huge quantity of post-harvest losses. Functioning of efficient marketing system is essential to make the farming operation remunerative. Prevailing agricultural marketing systems for these high-value crops like vegetables are not well-organized and the producers' share in consumer rupee is substantially low. The Economic Review (2009–2010) by Government of West Bengal has rightly highlighted that the, "horticulture sector of the state has been expanding fast and it has an enormous gainful employment potential. West Bengal has vast potential for development of food processing industries in the backdrop of its huge agro-horticultural raw materials. A growing market of high purchasing capacity, economically priced skill labor, self-sufficiency on power and availability of adequate water resources are the major forces behind the expansion of this area. With a view to boost up food processing industries the State Government has taken effective measures for minimizing wastage at all stages in the food processing chain by way of development of infrastructure for storage and transportation and also introduction of modern technology in these industries."

16.7 Livestock Income as a Source of Rural Livelihood

In the context of West Bengal, dairy has become more inclusive as compared to crop production in the sense that has involved majority of the vulnerable segments of the society for livelihoods. Nearly two-thirds of farm households in India have been associated with livestock production, and 80 % of them have been small landholders (less than 2 ha). It is also interesting to observe from Table 16.5 that about 96 % of dairy animals hold by the landless, small, and marginal farm households along with their largest share in land holdings. It clearly reflects how dairy animals

Table 16.6 Sources of income and expenses per month by the farmers during 2003. (Source: Authors' estimates from the unit level data on Situation Assessment Survey of Farmers, National Sample Survey Organisation (2005) 59th round, Ministry of Statistics and Programme Implementation, New Delhi)

Particulars	Landless	Marginal	Small	Medium	Large
Expenses (%)					
Off farm	6.49	0.16	0.04	0.01	0.01
Dairy farming	74.89	2.90	1.71	1.81	7.46
[a]Livestock farming	17.73	2.00	1.09	0.37	4.23
Crop	0.90	94.94	97.15	97.81	88.31
Total (₹)	127	4318	15,222	24,402	47,786
Receipts (%)					
Off farm	89.98	18.64	9.95	1.75	0.43
Dairy farming	6.87	2.15	1.44	1.18	1.23
[a]Livestock farming	3.02	0.79	0.55	0.04	0.53
Crop	0.13	78.42	88.06	97.03	97.81
Total (₹)	1481	8893	29,454	45,396	83,595

Landless 0–0.002 ha, *Marginal* 0.002–1.00 ha, *Small* 1.001–2.00 ha, *Medium* 2.001–4.00 ha, *Large* ≥4.00 ha

[a] Livestock farming excluded dairy farming

and land were integrated in the West Bengal context. Similarly, the households that were basically landless also had about 12 % of the dairy animals. It is also interesting to note that the distribution of dairy animals was far more uneven among the farm households than that of farmland suggesting that with efficient input and output support services, dairying can serve as a major economic activity for the small, marginal, and landless farmers.

In order to know the major sources of household level income, we have used the National Sample Survey Organisation 59th *Situation Assessment Survey of Farmers* (2002–2003) data. We have seen that there were four major sources of income (receipts) like crop, off-farm, dairy farming, and livestock farming (excluding dairy). Livestock keeping was one of the important sources of income for the landless and marginal farmers. From Table 16.6, it is clear that landless farm earned about 10 % of their income from livestock whereas about 7 % income was solely from dairy. It clearly reflects how dairy plays a vital role in case of landless farm households. If we compare the non-livestock keeping farm households to livestock keeping farm households, then we observe that majority of the non-livestock keeping households depended mainly on off-farm sources that were uncertain, time bound, and volatile in nature. Thanks to MGNAREGA scheme that helped to maintain the stability at household level income.

Another observation from Table 16.6 is that at the household level, landless farmers spend maximum amount for livestock keeping, whereas other farm households spend more on crop production. This is related to relative asset distribution across different types of productive assets and their uses, and it is the livestock including dairy animals that ensure higher stability in family incomes and hence relatively larger expenditure. Two important points emerge out of the 59th National Sample Survey Organisation data (2003) (i) Proportion of expenditure on livestock

Table 16.7 Decomposition of source-wise household level income inequality in West Bengal in 2003. (Source: National Sample Survey Organisation (2005) 59th round, Situation Assessment Survey of Farmers (2003) Ministry of Statistics and Programme Implementation, New Delhi)

Source	Income share	Gini of sources	Correlation with rank of total income	Percentage-contribution to total inequality	Source of inequality	Marginal effect
Off farm income	0.175	0.9526	0.8298	0.1787	0.1383	0.0037
Dairy farming receipts	0.0206	0.9061	0.663	0.016	0.0124	– 0.0046
Other livestock receipts except dairy	0.0075	0.981	0.6996	0.0066	0.0051	– 0.0008
Crop receipts	0.7969	0.801	0.9685	0.7986	0.6182	0.0018
Total income		0.774				

and dairying to total production expenditure of the households is inversely related to land ownership, and (ii) Proportion of household income from livestock and dairying to total family incomes of the households is also inversely related to land ownership. Essentially, the National Sample Survey Organisation data corroborates the significance of livestock and dairy incomes in the vulnerable segments of the farming community compared to land-owning classes of farmers.

16.8 Lessons from the Livestock Sector

In order to know the distribution pattern of the household income from different income sources, and to understand how those are correlated, source wise decomposition of Gini index was estimated by using Lerman and Yitzhaki (1985) method on the 59th National Sample Survey Organisation data on Situation Assessment Survey of Farmers (2002–2003). It was observed that 1 % incremental increase in crop income would trigger total income inequality by 0.2 % with a caveat that other things are unchanged. The result is expected to help in prioritizing the incorporation of different income sources to achieve higher social benefits and higher distributive justice. The Gini inequality of total household income is estimated at 0.774 (Table 16.7), indicating that income distribution among different household levels is unequal. The analysis further established that the share of crop income was highest (80 %) followed by off-farm income (18 %) and dairy farm income (2.06 %). On the other hand, the income from dairy farm and livestock source has a equalizing effect on the distribution of total income for all categories of farm households, which otherwise corroborates the hypothesis of relative income equalizing effect through dairying and livestock farming compared to distribution of incomes through crop.

This reemphasizes the importance of dairying in farming system for its dual beneficial social impact in improving incomes and reducing income inequality (Mandal et al. 2010). It also confirms that growth through inclusive dairying does not worsen income distribution, but helps in reducing absolute poverty. Therefore for production, aggregation in the dairy sector needs to (i) pace of innovation to enhance production and reduce costs (ii) reduce transaction costs, leveraging production of high value commodities (iii) expansion of cold-chain infrastructure, and (iv) ensure broader participation of societies, so that aggregate growth is large enough to accommodate landless, small, and marginal farmers.

16.9 Conclusions

Farming in West Bengal is individual-driven and unorganized. The average size of holding is only 0.82 ha which is much less than the national average (1.33 ha). Therefore, individual farmers, having very small marketable surplus of produce to sell in the market, will have to pay market price for all farm inputs and other basic utilities and consumable items. Thus, market access is not going to help majority of the West Bengal farmers. There is therefore, a need to organize West Bengal farmers on the model that helps the leadership to address the issue of organizing a vastly unorganized farming community, which help the farming community to gain from the market economy. The study concluded that in West Bengal, there is direct trade-off between the poverty reduction achievements and performance of agriculture in the state. The better performance during earlier period was mainly due to high growth in agriculture and during later period the agricultural performance slowed down and simultaneously the performance of the industry was also poor, thereby affected the overall performance in poverty reduction of the state. Smallholders are competitive in high-value agricultural activities, because of the availability of family labor and compete well in local markets. However, as production and marketing systems evolve, support to smallholders to provide efficient input services, links to output markets and risk mitigation measures will be important if they are to provide higher value products. Innovative public support and links to the private sector will be required for the poor to adapt and benefit as systems evolve.

The livestock, fishery, and horticultural sector could be an important pathway for increasing the income of the marginal and small farmers and landless laborers, as our result shows that the income from off-farm, dairy farm and livestock has an equalizing effect on the distribution of total income for all categories of farm households. This corroborates the hypothesis of relative income equalizing effect through dairying and livestock rearing compared to distribution of incomes through crop farming.

References

Banerjee A, Gertler V, Paul J, Maitresh G (2002) Empowerment and deficiency: tenancy reform in West Bengal. J Polit Econ 110(21):239–280

Bhattacharya M, Bhattacharya S (2007) Agrarian impasse in West Bengal in the liberalization era. Econ Polit Wkly 42(52):65–71 (29 December)

Devendra C, Thomas D, Jabbar MA, Zerbini E (2000) Improvement of livestock production in crop-animal systems in rainfed agro-ecological zones of South Asia. International Livestock Research Institute, Nairobi, p 108

Govt. of India (1972, 1982, 1992 & 2002) Agricultural statistics at a glance (Various issues), Department of Agriculture and Cooperation, Ministry of Agriculture, Government of India

Himansu (2007) Recent trend in poverty and inequality: some preliminary results. Econ Polit Wkly 42(6):497—508 (10th February)

Lerman RI, Yitzhaki S (1985) Income inequality effects by income source: a new approach and applications to the United States. Rev Econ Stat 67:151–156

Mandal S, Datta KK, Lama, TD (2010) Economic evolution of farming system research in NEH region: some issues. Indian J Agric Econ 65(1):118–134

National Horticulture Database (2008) Ministry of Agriculture, Govt. of India. http://www.nhb.gov.in. Accessed 23 Jan 2011

National Sample Survey Organisation (2005) Situation assessment survey of famers: 2003. National sample survey 59th round, ministry of statistics and programme implementation, Government of India

Paris TR (2002) Crop-animal systems in Asia: socioeconomic benefits and impacts on rural livelihoods. Agric Syst 71:147–168

Parthasarathy Rao P, Hall AJ (2003) Importance of crop residues in crop-livestock systems in India and farmers' perceptions of fodder quality in coarse cereals. Field Crops Res 84:189–198

Thomas D, Zerbini E, Parthasarathy Rao P, Vaidyanathan A (2002) Increasing animal productivity on small mixed farms in South Asia: a systems perspective. Agric Syst 71:41–57

Sengupta S, Gazdar H (1998) Agrarian politics and rural development in West Bengal. In Dreze J, Sen A (eds) Indian development: selected regional perspectives. Clarendon, Oxford, pp 129–204

Part III
Food and Livelihood Security through Agricultural Diversification

Chapter 17
Diversification and Food Security in the North Eastern States of India

K.C. Talukdar, Sima Das and Udeshna Talukdar

17.1 Introduction

North eastern region (NER) is lagging much behind the other parts of the country in farm diversification. The region experienced annual growth rate of 9.0 % in gross state domestic product (GSDP) and 7.51 % in per capita GSDP during the period from 2007–2008 to 2011–2012. The per capita annual GSDP in the region was ₹ 42,079, and the agricultural sector contributed about 30 % to GSDP. Agriculture is the main source of livelihood of about 85 % of population in the region. However, 40 % of rural population lives in poverty. Rice is the principal crop, and the crop sector contributes more than 60 % of total farm income. Farming is subsistence in nature and food grain is the main source of nutrients of the people.

Agricultural diversification is the means for increasing farm income and employment and it can play as the immediate sustainable source of food security through an integrated manner (Ram 1996; Chand 1996; Talukdar 2004). The region has wide biodiversity with more compatible flora and fauna, which can sustain with any climatic adversity in the region (Asati and Yadav 2004). Farm diversification also leads to nutritional security of the people (Birthal et al. 2006). The region is equally suitable to develop crops, livestock and fishery sectors in the states so as to meet the food and nutritional security and income of the rural people. It is hypothesized that farm diversification will increase the food and nutritional security in different states of the region. Attempt has been made to analyse changes in cropping pattern, crop and farm diversification, per capita food and nutritional security from domestic production in the states of north eastern India. The study was based on secondary

K.C. Talukdar (✉) · S. Das
Department of Agricultural Economics, Assam Agricultural University, Jorhat, Assam, India
e-mail: kctalukdar@yahoo.co.in

S. Das
e-mail: simaaau@gmail.com

U. Talukdar
Department of Economics and Sociology, Punjab Agricultural University, Ludhiana, Punjab, India
e-mail: udeshna_talukdar07@yahoo.com

M. Ghosh et al. (eds.), *Diversification of Agriculture in Eastern India*,
India Studies in Business and Economics, DOI 10.1007/978-81-322-1997-2_17

Table 17.1 Compound growth rate of area and extent of diversification of crop sector in north eastern states during 2001–2009

Crop groups	Assam	Manipur	Naga-land	Mizoram	Tripura	Arunachal Pradesh	Sikkim	Megha-laya	NER
Foodgrains	−7.02	14.53	17.71	−6.14	83.81	6.86	6.80	−0.30	14.53
Oilseeds	−17.17	5.56	51.24	–	5.97	12.56	−6.08	6.29	7.30
Vegetables	13.21	26.60	−41.31	28.85	−24.70	14.72	28.20	30.86	9.55
Fruits	212.45	53.15	983.00	59.00	−82.07	48.96	227.00	71.71	196.65
Simpson Index									
2001–2002	0.329	0.309	0.457	0.516	0.553	0.506	0.507	0.517	0.462
2005–2006	0.214	0.356	0.365	0.397	0.400	0.446	0.344	0.363	0.361
2009–2010	0.233	0.298	0.115	0.496	0.195	0.348	0.233	0.316	0.279

data published in different sources and tabular analysis was carried out to examine and interpret the parameters.

17.2 Results and Discussion

17.2.1 Crop Growth and Extent of Diversification

Area under the crops and its growth during the last decade and the change of diversification will indicate the shift in cropping pattern in the states of north eastern region. It was observed that in the states of Assam, Mizoram, Sikkim and Meghalaya, area under food grains decelerated during the last decade. Area under oilseeds in Assam also decelerated and it declined more in Assam. Area under rice is predominated in all the states to the extent of 80 % of Gross Cropped Area (GCA) as rice is the staple food in the region. Area under vegetables also decelerated in Nagaland and Tripura. Area under fruits was found to accelerate in all the states over years. This indicated that cropping pattern was shifted in the states towards high valued crops like fruits and vegetables. Introduction of National Horticultural Mission also enhanced the area under horticultural crops. In the north eastern region as a whole area under all crop groups accelerated with high growth in fruits (Table 17.1).

17.2.2 Horizontal Spread of Cropping Pattern

Proportionate area under the crops in gross cropped area will indicate the relative importance of the crops in the states of north east India. It was found that food grain crops predominated in the region. During 2001–2002 percentage of GCA varied from 66 % in Meghalaya to 82 % in Manipur followed by Assam. Area under food grains was predominantly higher in Assam, Manipur and Tripura. It was also observed that during 2009–2010 the proportionate area under food grains was found to increase in Assam (87.05%), Manipur (83.64%), Nagaland (93.95%), Tripura (89.05%), Arunachal Pradesh (77.52%), Sikkim (86.61%) and Meghalaya (80.28%) during the last decade. This indicated that the cropping pattern in the states of north eastern India was primarily food grains based. Area of oilseeds marginally increased in Sikkim and Meghalaya and indicated that relatively oilseeds area remained stagnant or declined.

17.2.3 Indices of Diversification

Diversification of crop sector was also examined in the states (Table 17.1) using the Simpson Index of diversification (Joshi 2001). This index was found to decline over years in Nagaland, Tripura, Sikkim and Meghalaya. This indicated that shift in area under the crops was erratic and diversification grew at low to moderate pace during the last decade. In the region Simpson Index was found as 0.279 which was quite low and the crop sector in the states was skewed towards specialization.

17.2.4 Diversification of Livestock Sector and Growth of Livestock Products

Livestock production like milk, eggs, meat and fish in different states of north eastern India and their growth indicated a higher shift of egg production in Tripura and Arunachal Pradesh and it grew faster in all states of the region except in Assam where it decelerated. Growth of meat production was also high in Tripura and Assam followed by Arunachal Pradesh. Production of meat in Nagaland over years during the last decade was found to decelerate. Fish production in the states of the region grew at a low pace. Examination of growth of livestock production in the region indicated that highest growth was for eggs and meat while fishery production grew at low rate. The region needs policy intervention for more production of milk and fish (Table 17.2). Meat production in Sikkim state was almost negligible.

During 2009–2010, the share of livestock sector was higher in Assam (69%), followed by Mizoram and Tripura. It was 7 and 6% in Nagaland and Arunachal Pradesh respectively and Sikkim shared 1% of total livestock in the region. In Assam 50% of meat is from pigs followed by 30% from goat meat followed by 19% of poultry meat.

Entropy index indicates the extent of farm diversification with integration of crops, livestock and fishery. Table 17.2 indicates that extent of diversification was higher in Sikkim and it was low to moderate in other states of the region. This was found mainly due to low development and growth of other sectors in the states. In all the states crop sector predominated and the total farm sector was inclined towards specialization and production of other sectors must be increased. The value of the index may further reduce if the fibre crops, spices and condiments are also considered.

Table 17.2 Compound growth rate of livestock products and extent of agricultural diversification in north eastern states during 2001–2009

Livestock products	Assam	Manipur	Nagaland	Tripura	Arunachal Pradesh	Sikkim	Meghalaya	NER
Milk	11.82	18.77	7.38	14.18	−39.40	26.17	20.73	8.52
Eggs	−6.19	45.01	42.30	152.03	372.25	48.34	8.10	93.40
Meat	76.18	10.67	−33.56	136.02	28.10	–	19.03	33.78
Fish	2.57	1.77	1.30	2.26	1.58	2.18	2.26	2.00
Entropy index of agricultural diversification								
2008–2009	0.321	0.310	0.300	0.300	0.323	0.583	0.30	0.310

17.2.5 Food Security, Concepts and Influential Factors

Food security is the economic access to food along with food production and food availability. It is a situation in which both food supply and effective demand are sufficient to cover nutritional requirements. Indicators of food security are household food availability, household food consumption and nutritional status. At household level food security refers to the ability of a household to secure adequate food to meet the dietary needs of all the members of the households (Kumar and Joshi 2013). Food security has been affected by growing urbanization, decrease of crop land, continuing crop loss, declining crop productivity and production, declining biodiversity, increase in population and change of taste and habits of the people. It happens when all people at all times have access to enough food that is affordable, safe and healthy, culturally acceptable, meets specific dietary needs, is obtainable in a dignified manner and is produced in ways that are environmentally sound and socially just.

17.2.6 Availability of Food from Domestic Crop Production

Per capita availability of food from domestic production in the states of north east India and the gap between normative requirement and availability indicates the food and nutritional status of the people. It was observed that per capita food grain availability in the states has marginally increased, but it was below the normal requirement in the states except in Nagaland in the recent years. It was found quite low in Mizoram and Meghalaya. In the entire region availability of food grains per capita per day was lower than normative requirement.

Availability of oil seeds (in terms of oil) per capita per day was found to decline in Assam, Mizoram and Tripura and it marginally increased in other states. It was more in Arunachal Pradesh, Sikkim and Nagaland and was found to increase in the region. In case of vegetables per capita availability was found to decline in Assam, Tripura and Nagaland. Availability was relatively higher in Assam, Sikkim, Meghalaya, Arunachal Pradesh and Tripura. In the entire region per capita per day availability of vegetables could meet the normative requirement. Fruit production in the region was also surplus in most of the states except in Arunachal Pradesh. The region witnessed a positive growth of all the food crops during the last one decade (Table 17.3).

Table 17.3 Per capita availability of food from domestic crop production in north eastern states of India (gms/cap/day)

Years	Assam				Manipur				Mizoram			
	Food grains	Oilseeds	Vegetables	Fruits	Food grains	Oilseeds	Vegetables	Fruits	Food grains	Oilseeds	Vegetables	Fruits
2001–2002	353.61	13.27	258.00	117.35	403.14	–	66.53	134.88	127.13	12.55	110.74	159.20
2005–2006	323.46	9.66	391.10	118.84	401.13	0.70	92.40	190.34	130.65	13.56	78.59	165.73
2009–2010	400.55	12.04	256.37	138.41	417.73	0.07	175.41	344.16	147.90	6.27	287.27	309.05
% change	1.47	−1.03	−0.07	2.00	0.40	–	18.18	17.24	1.82	−5.56	17.71	10.45
	Arunachal Pradesh				*Sikkim*				*Meghalaya*			
2001–2002	430.79	55.48	166.25	247.49	444.78	31.55	270.50	–	207.69	5.54	245.77	172.75
2005–2006	477.35	44.98	217.97	208.26	452.19	32.46	364.28	–	169.24	6.19	319.26	214.16
2009–2010	447.43	60.43	217.97	214.00	484.65	33.36	441.82	70.64	218.41	6.56	384.33	277.12
% change	0.43	1.00	3.46	−1.50	1.00	0.64	7.04	–	0.57	2.05	6.26	6.71
	Tripura				*Nagaland*				*Region*			
2001–2002	445.91	2.98	263.59	323.67	491.61	73.31	395.61	41.77	363.10	9.54	197.44	149.64
2005–2006	420.62	2.91	310.39	375.00	587.61	86.86	61.69	27.11	370.28	24.58	203.96	162.43
2009–2010	473.68	1.96	219.93	356.12	711.21	98.90	108.31	209.24	412.70	27.44	232.38	213.19
% change	0.69	−3.80	−16.56	1.11	4.96	3.88	−7.23	44.55	1.52	20.85	1.97	4.71

17.2.7 Availability of Food from Domestic Livestock Production

Per capita availability of livestock products like milk, eggs, meat and fish in the states of north eastern India was also examined and it was converted to nutrients availability in the states. It was observed that per capita per day availability of these products increased in the states except for eggs in Assam, milk in Arunachal Pradesh and Nagaland and fish in Meghalaya. Per capita availability of milk, eggs, meat and fish was higher in Sikkim, Meghalaya, Nagaland and Arunachal Pradesh and Tripura respectively. Meat and eggs were scarce in Sikkim. In the region, per capita per day availability of milk and meat in the region was higher than fish. However, level of meat and fish in the region had grown during the last decade (Table 17.4).

17.2.8 Nutritional Security from Plant and Animal Sources

This section shows the availability of nutrients like protein, minerals and calories from plant and animal sources produced domestically in the states of north east India. The gap between the availability and normative requirement will indicate the status of food and nutrients in the region. Availability of meat from domestic and free sources of forest sources also affects the food and nutritional security as the people are mostly nonvegetarian (Talukdar 2001). It was observed that protein from animal source increased more during the last decade relative to plant protein in the states. Level of animal protein was higher in Mizoram and Sikkim followed by Manipur. Out of total protein available share of plant protein was higher than animal protein in all the states of the region. Plant protein shared more than 90 % in Sikkim. It was further observed that level of total protein obtained from plant and animal sources was higher than normative requirement of protein. Protein surplus was more in Nagaland, Arunachal Pradesh and Manipur. Total protein was found deficit in Mizoram during 2001–2002 and 2005–2006 (Table 17.5).

17.2.9 Energy Requirement and Its Gap

Generation of energy in terms of kilocalories from domestic production and its gap between normative requirements was also examined in the states of north east India. It was observed that generation of energy was marginally higher than normative requirement per capita per day in several states like Tripura and Arunachal Pradesh and was found deficit in Assam, Manipur, Mizoram and Meghalaya. Level of energy in terms of kilocalories was found to have increased in Nagaland in the recent years. Decadal change over 2001–2002 indicated that energy in terms of kilocalories generated from domestic food sources decelerated in the states of Meghalaya, Arunachal Pradesh, Mizoram and Manipur over 2001–2002. Due to limitation of information, level of minerals and vitamins was not estimated in this paper. It is worth to mention that the people of north east states also take food from natural

Table 17.4 Per capita availability of food from domestic livestock production in north eastern states of India (ml/No/gms/cap/day)

Years	Assam				Manipur				Mizoram			
	Milk	Eggs	Meat	Fish	Milk	Eggs	Meat	Fish	Milk	Eggs	Meat	Fish
2001–2002	59.94	2.22	1.71	14.19	68.44	3.73.	16.46	16.55	35.15	3.62	19.36	7.91
2005–2006	65.66	2.35	2.37	16.52	77.50	4.20	23.15	18.34	37.66	4.09	22.60	9.41
2009–2010	66.18	2.04	2.72	17.52	78.51	5.56	23.15	18.92	42.68	5.16	32.64	9.44
PC change	10.41	−8.11	59.06	23.47	14.71	49.06	40.64	14.32	21.42	42.54	68.60	19.34
	Arunachal Pradesh				*Sikkim*				*Meghalaya*			
2001–2002	88.33	0.90	36.85	5.15	166.81	2.18	4.50	0.63	61.00	4.16	31.42	4.59
2005–2006	100.95	0.76	39.63	5.44	216.40	3.29	4.50	0.67	67.47	4.49	34.2	3.80
2009–2010	50.47	3.79	41.61	5.70	220.91	3.22	4.58	0.76	71.17	6.88	34.2	3.66
PC change	−42.86	321.11	12.92	10.68	32.43	47.71	1.78	20.63	16.67	65.38	8.85	−20.26
	Tripura				*Nagaland*				*Region*			
2001–2002	67.16	2.26	–	21.97	78.84	3.73	26.94	7.19	78.21	2.85	19.60	4.30
2005–2006	64.92	4.10	8.95	17.81	102.36	6.00	87.49	7.6	91.61	3.52	27.86	9.87
2009–2010	71.64	5.17	14.17	26.86	73.31	5.75	87.14	8.54	84.36	4.70	30.02	11.55
PC change	6.67	128.76	58.32	22.26	−7.01	54.15	223.46	18.78	7.86	64.71	53.16	168.60

Table 17.5 Per capita availability of food nutrients from domestic production in north eastern states of India (gm/cap/day)

Year	Assam									Manipur								
	Proteins			Minerals			Calories			Proteins			Minerals			Calories		
2001–2002	70.13 (91.36)	6.63 (8.64)	53.52	14.27 (93.33)	1.02 (6.67)	–	1575.10 (95.50)	74.26 (4.50)	−26.70	68.45 (86.13)	11.02 (13.87)	58.94	12.71 (90.33)	1.36 (9.66)	–	1552.88 (92.67)	122.86 (7.33)	−25.52
2005–2006	69.47 (90.93)	6.93 (9.07)	52.80	17.31 (93.52)	1.20 (6.48)	–	1567.95 (94.30)	94.70 (5.70)	−26.10	72.40 (84.40)	13.37 (15.60)	71.54	13.28 (89.37)	1.58 (10.63)	–	1625.60 (91.83)	144.56 (8.17)	−21.33
2009–2010	71.69 (90.00)	7.96 (10.00)	59.30	15.73 (92.75)	1.23 (7.25)	–	1627.65 (94.45)	95.61 (5.55)	−23.41	76.91 (84.85)	13.73 (15.15)	81.28	16.78 (91.20)	1.62 (8.80)	–	1960.63 (92.77)	152.78 (7.23)	−6.07
% change	2.22	20.06	10.80	10.23	20.60		3..34	28.77	−12.32	12.42	24.37	37.90	32.02	19.12		26.26	24.35	−76.21
	Mizoram									*Arunachal Pradesh*								
2001–2002	30.72 (79.13)	8.10 (20.87)	−22.36	45.98 (98.29)	0.80 (1.71)	–	760.75 (87.90)	104.76 (12.10)	−61.53	89.46 (87.60)	12.67 (12.40)	104.30	18.27 (93.55)	1.26 (6.45)	–	2126.12 (93.94)	137.18 (6.06)	5.27
2005–2006	28.36 (75.10)	9.40 (24.90)	−24.48	52.97 (98.31)	0.91 (1.69)	–	767.57 (88.86)	96.21 (11.14)	−61.61	96.18 (87.38)	13.89 (12.62)	120.14	19.58 (93.33)	1.40 (6.67)	–	2217.15 (93.60)	151.45 (6.40)	5.27
2009–2010	38.71 (76.44)	11.93 (23.56)	1.28	13.06 (92.50)	1.06 (7.50)	–	1145.96 (90.57)	119.29 (9.43)	−43.77	94.76 (88.28)	12.58 (11.72)	114.68	19.56 (94.86)	1.06 (5.14)	–	2203.73 (94.60)	126.01 (5.40)	3.54
% change	26.01	47.28	105.72	−71.60	32.50		50.63	13.85	−28.86	5.92	−0.71	9.95	7.06	−15.87		3.65	−8.28	−32.83
	Sikkim									*Meghalaya*								
2001–2002	87.97 (92.20)	7.44 (7.80)	90.82	17.14 (92.35)	1.42 (7.65)	–	1821.45 (91.96)	159.26 (8.04)	−11.97	45.23 (80.64)	10.86 (19.36)	12.18	7.75 (88.57)	1.00 (11.43)	–	1122.61 (90.25)	121.25 (9.75)	−44.72
2005–2006	92.80 (90.82)	9.38 99.18)	104.36	26.02 (93.42)	1.83 (6.57)	–	1925.34 (90.30)	206.78 (9.70)	−5.24	42.35 (78.59)	11.54 (21.41)	7.78	13.18 (92.56)	1.06 (7.44)	–	1053.86 (88.95)	130.93 (11.05)	−47.34
2009–2010	93.92 (90.77)	9.55 (9.23)	106.94	21.75 (92.12)	1.86 (7.88)	–	2098.11 (90.88)	210.41 (9.12)	2.60	53.34 (81.70)	11.95 (18.30)	30.58	16.38 (93.65)	1.11 (6.35)	–	1408.31 (90.63)	145.29 (9.36)	−30.95
% change	6.76	28.36	17.75	26.90	30.98		15.19	32.12	121.72	17.93	10.04	151.07	111.35	110.00		25.45	19.83	−30.80
	Tripura									*Nagaland*								
2001–2002	84.96 (88.92)	10.59 (11.08)	91.10	19.46 (93.02)	1.46 (6.98)	–	2091.74 (94.90)	112.56 (5.10)	2.03	109.45 (90.72)	11.20 (9.28)	141.30	22.91 (95.02)	1.20 (4.98)	–	2334.50 (94.64)	132.10 (5.36)	9.63

Table 17.5 (continued)

Year	Assam									Manipur								
	Proteins			Minerals			Calories			Proteins			Minerals			Calories		
2005–2006	83.14 (89.63)	9.62 (10.37)	85.52	20.41 (93.94)	1.30 (6.06)	–	2112.88 (94.90)	113.29 (5.10)	1.06	115.23 (81.67)	25.87 (18.33)	182.20	18.48 (90.15)	2.02 (9.85)	–	1909.73 (88.60)	245.92 (11.40)	−4.19
2009–2010	87.86 (86.81)	13.35 (13.19)	100.26	19.4 (91.25)	1.86 (8.75)	–	2182.63 (93.78)	144.62 (6.21)	3.43	141.21 (85.04)	24.84 (14.96)	232.10	24.77 (93.08)	1.84 (6.92)	–	3156.76 (93.46)	220.93 (6.54)	50.12
% change	3.41	26.06	10.05	−0.31	27.40		4.34	28.48	68.96	29.02	121.78	64.26	8.19	53.33		35.22	74.81	420.46

1 = Plant source, 2 = Animal source, 3 = Surplus/deficit (%)
Figures in parentheses indicate percentage share

plant sources (local herbs) collected freely and animals hunted from the forest (Talukdar 2001). Food and nutritional status in the hill states and also in the plains to some extent will be improved if foods obtained from natural sources are taken into consideration. It is imperative that food security of the people will be affected by its accessibility, economic status, etc.

17.3 Conclusion

The paper has indicated that although the region has gained area in growth of the crops, the crop diversification in the region was quite low and the crop sector was almost specialized. The region has witnessed high growth of eggs and meat and was higher in Tripura. Low level of entropy index showed a low pace of agricultural diversification in the region. There was a deficit of food grains and edible oil with sufficient level of fruits and vegetables. The region must be capitalized to upkeep the livestock sector for meeting the nutrient deficiency. This will also enhance diversification to increase income, employment and meeting food security in the region. The region needs long term strategy to enhance the domestic production of crops and livestock with more diversification of farm sector for meeting the food and nutrient deficiencies in several states of the region after meeting the constraints.

References

Anonymous, Statistical Handbook of Assam, Directorate of Economics and Statistics, Government of Assam, various published issues. Accessed 8–11 May 2011

Anonymous, Basic Statistics of NE Region, NEC Secretariat, Shillong, various published issues. Accessed 6, 20, 25 May 2011

Asati BS, Yadav DS (2004) Diversity of horticultural crops in North Eastern region. Himal Ecol 12:1–11

Birthal PS, Jha AK, Joshi PK, Singh DK (2006) Agricultural diversification in North Eastern region of India: implications for growth and equity. Indian J Agric Econ 61(3):469–480

Chand R (1996) Diversification of Punjab agriculture: need and implications for food security. Agric Econ Res Rev 9(2):37–44

Joshi PK (2001) Crop diversification in rainfed agriculture. Agricultural Economics Research Review, Conference Proceedings

Kumar P, Joshi PK (2013) Household consumption pattern and nutritional security among poor rural households: Impact of MGNERGA. Agric Econ Res Rev 26(1):73–87

Ram GS (1996) Food security system, poverty and rural development. Agric Econ Res Rev 9(2):121–127

Talukdar KC (2001) Study of socio-economic condition and food security status of tribals of North East India (case of Nagaland and Mizoram). Jai Vigyan National Science and Technology Mission on Household Food and Nutritional Security, ICAR, Department of Agricultural Economics, Assam Agricultural University, Jorhat

Talukdar KC (2004) Prospects and determinants of crop diversification with high valued crops in marginal and small farms in the flood affected areas of Assam. ICAR ad-hoc project, Department of Agricultural Economics, Assam Agricultural University, Jorhat

Chapter 18
Impact of Improved Agro-Techniques Towards Food and Livelihood Security

S. C. Sarker, P. S. Patra, G. Mula and B. Paramanik

18.1 Introduction

Though India has substantially increased the production and productivity of food crops and raised its rural incomes, the country still has a large number of poor and malnourished populations. Moreover, Indian agriculture is currently facing greater challenges from unsustainable use of natural resources, deceleration in the growth of agriculture and from threats as well as opportunities of opening of agricultural markets. There are wide gaps in yield potential and national average yields of most crop enterprises (Sulaiman 2008). Sustainability in food production, inclusive growth and gender mainstreaming, enhancing rural employability, harnessing full production potential of all regions with required diversification, strategic manoeuvring of growing biotic and abiotic stresses, and appropriate exploitation of natural resource opportunities are few other vital issues receiving national priority.

At present, there are only two ways to satisfy the increasing food and other agricultural demands of the country's rising population: either expanding the net area under cultivation or intensifying cropping over the existing area. The net sown area

S. C. Sarker (✉)
Department of Agricultural Economics, Uttar Banga Krishi Viswavidyalaya, Pundibari, Cooch Behar, West Bengal, India
e-mail: sarker57@rediffmail.com

P. S. Patra
NAIP, Component-3, Department of Agronomy, Uttar Banga Krishi Viswavidyalaya, Pundibari, Cooch Behar, West Bengal, India
e-mail: parthaagro@gmail.com

G. Mula
NAIP, Component-3, Department of Agricultural Economics, Uttar Banga Krishi Viswavidyalaya, Pundibari, Cooch Behar, West Bengal, India
e-mail: gobindamula@gmail.com

B. Paramanik
NAIP, Component-3, Department Soil Science and Agril. Chemistry, Uttar Banga Krishi Viswavidyalaya, Pundibari, Cooch Behar, West Bengal, India
e-mail: paramanikbappa@gmail.com

M. Ghosh et al. (eds.), *Diversification of Agriculture in Eastern India*,
India Studies in Business and Economics, DOI 10.1007/978-81-322-1997-2_18

of the country has risen by about 20 % since independence and has reached a point where it is not possible to make any appreciable increase. Thus, raising the cropping intensity is the only viable option left.

The introduction of new cultivated species and improved varieties of crop is a technology aimed at enhancing plant productivity, quality, health and nutritional value and/or building crop resilience to diseases, pest organisms and environmental stresses. Crop diversification refers to the addition of new crops or cropping systems to agricultural production on a particular farm taking into account the different returns from value-added crops with complementary marketing opportunities. Major driving forces for crop diversification include: increasing income on small farm holdings, withstanding price fluctuation, mitigating effects of increasing climate variability, balancing food demand, improving fodder for livestock animals, conservation of natural resources, minimizing environmental pollution, reducing dependence on off-farm inputs, depending on crop rotation, decreasing insect pests, diseases and weed problems and increasing community food security.

A sustained economic growth, rising per capita income and growing urbanization are ostensibly causing a shift in the consumption patterns in favour of high-value food commodities like fruits, vegetables, dairy, poultry, meat and fish products from staple food such as rice, wheat and coarse cereals. The demand for and supply of these commodities have grown much faster than those of food grains (Kumar et al. 2003; Joshi et al. 2004). The share of these commodities in the total expenditure on food increased from 34 % in 1983 to 44 % in 1999–2000 in the rural areas, and from 55 to 63 % in the urban areas (Kumar and Mruthyunjaya 2002). This change is not only confined to the higher income group of the Indian society only but also visible in the lower income or 'below poverty line' segment. Such a shift in consumption patterns in favour of high-value food commodities even among the poorest strata of the Indian society depicts an on-going process of transformation that is leading towards a 'silent revolution' of agricultural diversification.

Agricultural diversification in India is slowly picking up momentum in favour of high-value food commodities primarily to augment income rather than the traditional concept of risk management. The nature of diversification differs across regions due to existence of wide heterogeneity in agro-climatic and socioeconomic environments. It was considered interesting to delineate the key regions and subsectors of agriculture where diversification was catching up fast. Keeping the above background the present study was conducted to address the impact of improved agro techniques towards food and livelihood security.

18.2 Methodology and Database

18.2.1 Database

A critical appraisal of study on impact of improved agro techniques towards food and livelihood security furnished a good conceptual as well as structural platform for the present investigation and assisted in using of suitable methodology. In this part, profile of the study area, data sources and the analytical tools were used to address the specific objectives.

The present study was carried out on the basis of both secondary and collected primary research data through field experimentation from the beneficiaries of NAIP-3 running since 2008 covering the period up to December 2011. The sample units were scattered over ten villages of Itahar, Tapan, Manickchalk and Suti-I Block of Uttar Dinajpur, Dakshin Dinajpur, Malda and Murshidabad district of West Bengal, India respectively. The sample frame composed of 1314 beneficiaries included under the 'sustainable livelihood empowerment sub-project' of National Agricultural Innovation Project.

Four backward districts of the state West Bengal namely Uttar Dinajpur, Dakshin Dinajpur, Malda and Murshidabad were brought under sustainable livelihood empowerment sub-project of National Agricultural Innovation Project. The backwardness is characterized by low per capita income, low yield per acre of land, backwardness in industrialization, shortage of capital and entrepreneurship, and the lack of infrastructure and large labour surplus. The West Bengal Human Development Report (2004) is clearly indicative of the prevalence of very poor health index, gender development index, high degree of infant mortality rate (IMR) and low human development index (HDI) values across all these districts. A highly semi-humid atmosphere and medium rains characterized the climate of these districts. The average annual rainfall 1487.32 mm and January to February is the coldest period with temperature varying between 7.4 to 20.3 °C, April is the hottest month with mean daily maximum of 38.5 °C and mean daily minimum of 22.2 °C. The economy of these backward areas is mostly dependent upon the primitive agriculture and its allied activities. The total agricultural lands of these four districts are 86.41, 79.29, 76.09 and 76.15 % respectively (Census 2001). The districts have abundance of natural resources like bamboo, palms, creepers, fruit trees etc; agricultural produce like rice, jute, potato, wheat and maize and mustard in some pockets. So, by making proper planning like introducing short gestation vegetables, improved varieties of field crops, utilizing homestead area, creation of small-scale village handicrafts through self help groups (SHGs), there is a wide scope for the development of these districts.

An extensive survey instrument has been designed for the purpose of conducting the Baseline survey. The instrument had been pre-tested and fine tuned on the basis of pre-testing feedback. The survey instrument has two components such as questionnaire based and PRAs and Focus Group Discussion. Data collected through Questionnaire Based Survey was triangulated by the PRA outputs and interview results. Besides that a sample check had also been conducted to estimate the extent of data sanitization. Field investigators had primarily been selected from the same locality living in close vicinity of the identified villages. The selection criteria set to choose the investigator is knowledge of conducting surveys, thorough understanding on the local areas. The Program Coordinators of respective Krishi Vigyan Kendra (KVKs) are involved in selection process.

To accomplish the objective of this investigation, data on yield, net sown area and cost of cultivation were determined from the experimental fields from the clusters of the study areas. Here, cluster represents the groups of adjacent villages, selected for the study. The information was also collected from government officials and staff in agricultural agencies for cross verification.

18.3 Analytical Tool

Cropping diversification: One of the goals of sustainable agriculture is to minimize the farmer's risk. Cropping diversification can reflect this point due to the fact that one crop failed can be compensated from another crop. To see the differences in crop diversification among the study area, the following formula was used from Thapa (1990).

$$\text{ICD}(\%) = [\text{Pa} + \text{Pb} + \text{Pc} + \ldots + \text{Pn}]/\text{Nc}$$

where

ICD	Index of Crop Diversification
Pa	Proportion of sown area under crop a (%)
Pb	Proportion of sown area under crop b (%)
Pc	Proportion of sown area under crop c (%)
Pn	Proportion of sown area under crop n (%)
Nc	Number of crops

Cropping Intensity = (Gross cropped area/Net sown area) · 100

18.4 Results and Discussions

Cropping diversification was used to explain the risk of farming. The value of index starts from 100 (when only one crop is grown) and goes to the tendency of zero (when as many as crops are grown). The lesser the index value is, the higher the cropping diversification will be, and thus, more relatively sustainable for the farming system. The cropping diversification index (ICD) was enhanced from 36.90 to 18.73 %, 40.03 to 19.72 %, 34.04 to 19.90 % and 43.71 to 22.33 % at cluster-I, cluster-II, cluster-III and cluster-IV, respectively (Table 18.1). This was due to more area being brought under cultivation, introduction of high-yielding short-duration varieties and substitution of rice, jute and potato growing area to some extent with other profitable field crops like groundnut, maize, lentil, green gram and black gram.

Data presented in Table 18.2 revealed that among the crops, highest percentage of productivity was achieved in black gram cultivation (87.5) followed by maize (58.45), green gram (56.25) and rice (54.93). The reason might be the introduction of short-duration high-yielding varieties namely Sarada (WBU-108), Ganga-1, Sonali and IET-4786 and improved agro–techniques which led to an increase in the net income and accelerated the food security. As cumulative effect of integration of improved crop husbandry, new crop sequencing and replacement of varieties, impressive increase in productivity has been achieved for all major cereals (49.63 %), oilseeds (43.02 %), and pulses (58.44 %) as well as potato (23.85 %) to effectively address food insecurity issue in the study area.

Zero-tillage practices on an average produced 7.86 % more grain yield than conventional tillage method. The probable reason might be delayed sowing, which

Table 18.1 Analysis of crop diversification index (ICD) before and after NAIP intervention

Name of the crops	Net sown area (ha)							
	Cluster-I		Cluster-II		Cluster-III		Cluster-IV	
	Before	After	Before	After	Before	After	Before	After
Rice	78.66	69.23	79.81	71.00	74.01	66.50	84.15	69.25
Jute	47.21	44.33	51.45	41.35	55.72	47.75	58.50	55.23
Potato	34.43	26.72	31.90	28.56	36.05	31.05	38.08	34.50
Wheat	21.28	28.25	23.75	31.50	19.00	23.74	14.79	21.81
Mustard	–	9.25	13.22	10.48	15.27	9.33	23.01	16.33
Lentil	–	7.27	–	7.98	–	6.58	–	9.78
Black gram	–	2.93	–	5.44	4.17	13.25	–	3.22
Groundnut	–	5.23	–	6.52	–	7.19	–	10.42
Green gram	–	3.12	–	4.05	–	4.95	–	13.11
Maize	2.94	7.25	–	8.26	–	2.75	–	6.37
Sesame	–	2.50	–	1.74	–	3.86	–	5.59
Total sown area (ha)	298.20	298.20	185.00	185.00	183.00	183.00	798.00	798.00
ICD	36.90	18.73	40.03	19.72	34.04	19.90	43.71	22.33

Table 18.2 Productivity gain through replacement of varieties and improved agro-techniques

Crops		Area covered in hectare	Number of household involved	Productivity ($t\ ha^{-1}$)		% gain
				Initial	Improved	
Cereals	Rice	105.56	748	1.42	2.2	54.93
	Wheat	115.96	603	2.14	2.90	35.51
	Maize	60.00	600	2.84	4.50	58.45
Pulses	Black gram	21.38	180	0.400	0.750	87.50
	Green gram	10.28	150	0.560	0.875	56.25
	Lentil	9.36	120	0.950	1.25	31.58
Oilseeds	Sesame	49.92	450	0.382	0.565	47.91
	Mustard	115.04	750	0.670	0.90	34.33
	Groundnut	10.16	175	1.26	1.85	46.83

leads to less time being available for the physiological growth and development and high temperature, which triggers the grain filling stage (Bear et al. 1994) under conventional tillage. The yield of wheat in zero-tillage practice was more than the conventional tillage which could be attributed to better utilization of soil moisture, water-use efficiency, nutrient uptake and less fluctuation of soil temperature (Cham et al. 2002 and Bauer et al. 2002).

Total cost in wheat cultivation was varied among the clusters due to varied cost of labour and hiring machineries. Mean data showed that zero-tillage field recorded higher gross and net return to the tune of ₹ 35,167.75 and 16,890.21 ha^{-1} (Table 18.3) than conventional tillage where gross and net return was ₹ 32,641.12 and 9907.29 ha^{-1}. The probable reason is comparatively higher grain yield and lower cost of cultivation. From the mean data of four clusters it was observed that zero-tillage method save total cost of ₹ 4456.30 ha^{-1} and increase profit margin of ₹ 6982.90 ha^{-1}.

Table 18.3 Yield and economics of wheat cultivation under conventional and zero-tillage practice (Returns ₹/ha, yield t/ha)

Clusters	Conventional tillage practice					Zero-tillage practice				
	Grain yield	Cost of cultivation	Gross return	Net return	B:C ratio	Grain yield	Cost of cultivation	Gross return	Net return	B:C ratio
Cluster-I	3.12	22434.70	32003.90	9569.19	0.43	3.45	18475.91	35314.67	16838.75	0.91
Cluster-II	3.26	23478.27	33452.58	9974.32	0.42	3.51	18590.72	35994.58	17403.87	0.94
Cluster-III	2.95	20947.75	30232.38	9284.63	0.44	3.15	17117.79	32256.75	15138.96	0.88
Cluster-IV	3.40	24074.63	34875.63	10801.00	0.45	3.62	18925.75	37105	18179.25	0.96
Mean	3.18	22733.84	32641.12	9907.29	0.44	3.43	18277.54	35167.75	16890.21	0.92

Zero-till field also fetched higher benefit-cost (B-C) ratio than conventional tillage irrespective of clusters. Mean data of Table 18.3 exhibited that zero-till field recoded higher B-C ratio of 0.92 as compared to conventional tillage method (0.44). This was simply due to higher net return and lower cost of cultivation.

Higher cropping intensity means that a higher portion of the net area is being cropped more than once during one agricultural year. This also implies higher productivity per unit of arable land during one agricultural year. Gross cropped area enhanced by 42.14, 34.72, 26.58 and 14.25 % at cluster-II, cluster-III, cluster-I and cluster-IV respectively, after the introduction technological interventions through National Agricultural Innovation Project (NAIP) subprojects in the study areas. Although net sown area was found highest at cluster-IV and cluster-I, than rest of the clusters, but utilization of net sown area was comparatively less due to low adoption of introduced technologies and large land holding size.

Intensification of cropping was revealed highest at cluster-II (217.03 %) followed by cluster-I (191.13 %), cluster-III (178.28 %) and cluster-IV (158.52 %) which was 64.34, 40.13, 45.95 and 19.77 % higher over their respective baseline value due to interventions on new crop sequencing, resource conserving technologies and varietal replacement.

18.5 Conclusions

The present chapter is an outcome of consortium based pluralistic action research endeavour under component-3 of NAIP. The research was purposed for bringing about change in the quality of life through widening the option basket for target communities, required knowledge and choice facilitation to them. Among the introduced improved agro techniques zero-tillage wheat cultivation, varietal replacement and new crop sequencing gained popularity, and adaptability increased the profit margin, crop diversification and cropping intensity, which led to uplift the socioeconomic condition and promote living with food and nutritional security in the study areas.

References

Bauer PJ, Frederick JR, Busscher WJ, Van SE (2002) Optimizing conservation tillage production: soil specific effects of management practices on cotton, soybean and wheat. Proceedings of 25th annual southern conservation tillage conference for sustainable agriculture. USA, pp 383–385

Bear MH, Hendrix PF, Coleman DC (1994) Water stable aggregates and organic carbon fractions in conventional and no tillage soils. Soil Sci Am J 58:777–786

Census of India (2001) http://westbengal.gov.in/portal/banglarMukh/CensusPage/CensusPortletWindow;jsessionid=EB78A0817434FAEEB2B035FE1594758A.node1?action=e&windowstate=normal&mode=view

Cham KY, Heenan DP, Gates A (2002) Soil carbon fractions and relationship to quality under different tillage and stubble management. Soil Tillage Res 63:133–139

Joshi PK, Gulati A, Birthal PS, Tewari L (2004) Agricultural diversification in South Asia: patterns, determinants and policy implications. Economic and political weekly 39(24) June 12–18, pp 2457–2468

Kumar P, Mruthyunjaya (2002) Long term changes in food basket in India. Paper presented in an International workshop on agricultural diversification in South Asia, jointly organized by MoA, Bhutan, NCAP and IFPRI, Paro, Bhutan, 21–23 Nov 2002

Kumar P, Mruthyunjaya, Birthal PS (2003) Changing consumption pattern in South Asia. Paper presented in the International Workshop on Agricultural diversification and vertical integration in South Asia organized by FICCI-ICRISATIFPRI on 5–6 Nov 2003 in New Delhi

Sulaiman RV (2008) Knowledge generation and management. In: Ghosh GN, Ganguly R (eds) Development challenges of Indian agriculture: background technical papers for the preparation of the national medium term priority framework for FAO and the Government of India, Information Division, FAO, Rome, Italy

Thapa GB (1990) Integrated watershed management in Pokhana valley in Nepal. Ph.D. dissertation, school of environment, resources and development. Asian Institute of Technology, Thailand

Chapter 19
Floriculture in West Bengal in Augmenting Income and Export

T. N. Roy, K. K. Das and D. Rai

19.1 Introduction

The value and essence of flowers have been increasing day by day for its multipurpose utilities and its emerging economic value under trade liberalization and farm diversification. After liberalization, floriculture has been identified as a sunrise industry and accorded 100% export-oriented status (Anonymous 2009). With the increase of per capita income and change of consumer preferences, it appears that there is a steady increase in demand of flower, and thus, floriculture has become one of the important commercial trades in agriculture. India experiences 25% annual growth in its domestic market. As a result, commercial floriculture has emerged as a farm operation with scientific recommendations as well as hi tech activity taking place under controlled climatic conditions inside greenhouses. Floriculture products mainly consist of cut flowers, pot plants, cut foilage, seeds bulbs, tubers, rooted cuttings, dried flowers or leaves and other value-added products. The important floricultural crops in the international cut flower trade are rose, carnation, chrysanthemum, garbera, gladiolus, gypsophila, liastris, nerine, orchids, archilea, anthuriu, tulip, lilies, etc. (Anonymous 2000).

The growth of agriculture is critically dependent upon value-addition and processing. Value-added floriculture is the process of increasing economic value and consumer appeal for a floricultural commodity. Increase in value and appeal of floricultural product or commodity are done through changes in genetics, processing or diversification which requires more time, labour and skill than the typical farming operations. The value-addition for marketing flowers includes adoption of post-harvest technology and improved logistics. In the era of

T. N. Roy (✉) · K. K. Das · D. Rai
Department of Agricultural Economics, Uttar Banga Krishi Viswavidyalaya, Pundibari, Coochbehar 736165, West Bengal, India
e-mail: tnr_ubkv@yahoo.co.in

K. K. Das
e-mail: kkdas_ubkv@yahoo.com

M. Ghosh et al. (eds.), *Diversification of Agriculture in Eastern India,*
India Studies in Business and Economics, DOI 10.1007/978-81-322-1997-2_19

Table 19.1 Area and production of flower in India 2000–2010. (Source: National Horticulture Board (NHB) 2010, obtained from www.nhb.org.in on *10/10/2011*)

Year	Area (thousands of hectares)	Production	
		Loose (thousands of megatons)	Cut (millions)
2000–2001	98	556	803
2001–2002	106	535	2565
2002–2003	70	735	2060
2003–2004	101	580	1793
2004–2005	116	655	1952
2005–2006	129	654	2920
2006–2007	144	800	3716
2007–2008	161	870	4342
2008–2009	167	987.4	47,942
2009–2010	175.72	1016.65	49,082

trade-oriented horticulture, value addition is used for reasons like (i) unstable prices of raw materials, (ii) federal farm policy, (iii) changing consumer preferences, (iv) making more money by cutting out the middlemen, etc. In fact, one of the most important aspects of floriculture marketing is to give customers a reason to buy floricultural (value-added) products.

The industry is mostly characterized by the sale of mostly loose flowers (rose, chrysanthemum, jasmine, marigold, crossandra, tuberose, etc.) and cut flowers (rose, carnation, chrysanthemum, gladiolus, gerbera, orchids, anthurium, liliums, alstroemeria, tulip, etc.). India is second in loose flower production after China. Presently, the world's total flower foliage area is 855,380 ha and the total area under protected cultivation is 346,663 ha (APEDA 2011). In India 175,720 ha area is under floriculture, producing 1016.65 MT of loose flowers in the year 2009–2010. Cut flowers have increased over the years to touch a production of 49,082.00 million flowers in 2009–2010. The industry has been growing at a Compounded Annual Growth Rate (CAGR) of 25 % over the past decade, with production area growing at a CAGR of 6.89 % since 2000–2001. Table 19.1 shows the present production status of floriculture in India.

Floriculture is an emerging agro-based industry in West Bengal. It has become a highly profitable economic activity due to varied and favourable agro-climatic conditions. The major flowers produced are (i) rose, (ii) tuberose, (iii) gladiolus, (iv) gerbera, (v) carnation, (vi) cockscomb. Besides in hilly areas, particularly in Darjeeling District, some varieties of orchid are also being commercially produced. Flowers in West Bengal are mainly grown in places like Kalimpong, Panskura, Ranaghat, Thakurnagar and Bagnan in addition to other places around the state. West Bengal, with its excellent logistics, research and extension support in floriculture trade, has huge potential to capture the booming export markets of Europe and Japan. Presently, the State produces 22,170 million cut flowers, for which the State ranks first by contributing approximately 5.5 % of country's total production during 2009–2010 (Anonymous 2010). In the case of loose flowers, total production stands at 55.177 MT, which occupies fifth position, sharing nearly 43.68 % of country's total production during the same period (Anonymous 2010).

The West Bengal government's department has advised farmers to examine ways to take up the making of flower-based products, as the demand for basic flowers was falling due to their seasonal production. These products were expected to help farmers avoid losses or go in for distress sale of surplus flowers at rock-bottom prices. Farmers in many places earn their livelihood by adopting floriculture commercially. For example, nearly 25,000 of people of Medinipur district are directly or indirectly involved in commercial flower production.

In view of the above, this study analyses the scope and prospect of value addition of flowers in West Bengal with some specific objectives. These are (1) scope and prospect for producing value-added products, (2) essential requirements and available infrastructures for the industry, (3) marketing facilities and mechanism (domestic and export), (4) constraint and suggestion thereupon for development of the industry.

19.2 Results and Discussion

19.2.1 Scope and Prospect

India possesses some unique and favourable situations for producing different types of flowers like (i) varied agro-climatic conditions, (ii) abundant sunlight and water, (iii) availability of skilled and semi-skilled labour, (iv) proximity to West Asia, Southeast Asia, the Far East and Europe, (v) insight into international markets gained in the last decade, (vi) a large and rapidly growing domestic market, (v) government support for export-oriented production, (vii) cost-effective technology for protected cultivation and (viii) upcoming handling facilities at major airports.

Of late, India has developed some standardized technologies for the improvement of the floriculture industry like (i) Tissue culture for mass scale production, for example of orchid, carnation and lilium; (ii) foliage plants; (iii) automated green (poly) houses—rose, carnation; (iv) shade net—Gerbera, anthurium; (v) cool chain and post-harvest management—rose, carnation, gerbera; (vi) Breeding of anthurium: (vii) production of artificial media like coco-peat, peat moss and vermiculite. Table 19.2 shows the loose flower production in major producing states in India. Table 19.3 shows the cut flower production in major producing states in India, which reveals that West Bengal occupies the first position with respect to the production of cut flowers in India.

Table 19.4 shows the potential areas of flower industries in India. The industry is concentrated mainly in Himalayan and sub-Himalayan areas and southern India. In West Bengal, the flower industry is located in Kolkata, Siliguri, Darjeeling and Kalimpong. In Eastern Himalayas areas, particularly in Darjeeling district, rhododendron remained a untapped flower for commercial purpose.

Table 19.2 Loose flower production in India 2010. (Source: Anonymous 2010a)

State	Area (thousands of hectares)	Percentage of total area	Production (thousands of megatons)	Percentage of total production
Tamil Nadu	32	18.197	247.3	24.325
Karnataka	27	15.354	203.9	20.056
Andhra Pradesh	19.5	11.089	125	12.295
Maharashtra	17.5	9.951	91.1	8.961
Punjab	1.7	0.967	82	8.066
West Bengal	21.9	12.453	55.177	5.427
Haryana	5.5	3.128	53.9	5.302
Gujarat	12.5	7.108	49.5	4.869
Orissa	7.6	4.322	32.2	3.167
Jharkhand	1.6	0.91	22	2.164
Uttar Pradesh	10.4	5.914	17.6	1.731
Total	175.72	–	1016.65	–

Table 19.3 Cut flower production in India 2010. (Source: Anonymous 2010a)

States	Production (thousands of megatons)	Percentage of total production
West Bengal	22,170.00	43.677
Maharashtra	7914.00	15.591
Karnataka	5860.00	11.545
Gujarat	5063.00	9.975
Uttar Pradesh	2958.00	5.828
Uttarakhand	2056.00	4.051
Jharkhand	1711.00	3.371
Delhi	1038.00	2.045
Haryana	929	1.83
Himachal Pradesh	566	1.115
Total	50,759.00	100

Horticultural crops (products) in general, and floriculture in particular, provide an excellent opportunity for diversification which paves the way to produce different demand-driven, value-added products. These diversified products are also capable of reducing investors' risk in agri-business and also help augment the export potential of high-value crops, as being witnessed after trade liberalization under World Trade Organization (WTO) agreements.

Presently, India's floriculture industry produces diversified products which are (1) cut flowers—rose, carnation, gladiolus, gerbera, etc.; (2) novelty flowers—anthurium, orchid, sterlitzia (bird of paradise), heliconia, etc.; (3) potted plants; (4) ornamentals; (5) foliage and (6) planting materials—seeds, bulbs, cuttings, etc. Among the above-mentioned flowers, floriculturists of West Bengal continue to invest in the production of rose, gerbera, gladiolus, anthurium, orchid, tuberose, marigold, etc.

Table 19.4 Potential areas of flower industries in India. (Source: APEDA 2011; www.apeda.gov.in)

Crop	Place
Rose	Pune, Bangalore, Nasik
Carnation	Pune, Bangalore, Ooty, Kodaikanal, Solan, Shimla, Nasik, Palampur
Chrysanthemum	Bangalore, Pune, Kolkata, Shimla, Solan
Anthurium	Coorg, Trivandrum, Trichur, Siliguri
Gerbera	Pune, Bangalore, Guwahati, Kalimpong, Coimbatore
Orchid	Kalimpong, Darjeeling, Mirik, Gangtok, Shimla, Srinagar, Shillong, Mumbai, Kolkata, Chennai, Siliguri, Trivandrum, Trichur,
Lilium	Shimla, Solan, Ooty, Nainital, Kalimpong, Darjeeling
Tulip	Palampur, Srinagar
Alstroemeria	Shimla, Srinagar, Ooty, Kalimpong
Iris	Solan, Shimla, Kalimpong

19.2.2 Areas of Value-Addition in Floriculture

Value-added products are the need of the time to sustain this industry from stiff market competition. Value addition to the flower follows some processing operations. The processing industry is becoming the key for modern farm business as (i) development and growth of agriculture is critically dependent upon value addition and processing and (ii) increased export will motivate farmers to achieve better productivity besides inducing entrepreneurship. If appropriate infrastructural facilities are provided, rural unemployed people can be absorbed in different value chain segments. This effort may help reduce rural unemployment to some extent. Value-addition to floriculture would be effective when it fulfils some conditions. These are: (1) a value-added marketing approach is adopted to extend the spread of the marketing, (2) value-addition ensures high premium to the grower while providing more acceptable quality products for the domestic and export markets, (3) the value-addition for marketing flowers includes adoption of post-harvest technology and improved logistics. The broad areas of value-added floriculture are (1) essential oil extraction from flowers, (2) cut flower production for occasion/flower forcing (fresh and dried), (3) live plants and potted plants and (4) foliage and other parts of plants (fresh/dried/potpourri). The following are some value-added products:

(i) Potpourri: The art of preserving the beauty, colour and the fragrance of flowers can take any number of creative forms—from dried flower blossoms which are then used in wreaths or unusual table decorations to the creation of a flower. Potpourri is a mixture of dried, sweet-scented plant parts including flowers, leaves, seeds, stems and roots. The basis of a potpourri is the aromatic oils found within the plant. A significant component of dry flower is also an important component of export.

(ii) Essential oils: Essential oils and perfumery made from natural sources are in great demand and available from the flower crops which include mainly rose, jasmine, tuberose, etc.
(iii) Natural dyes: Natural pigments and colour can be extracted for commercial uses. Marigold pigments are widely used in the poultry industry to enhance the colour of the meat and yolk of the eggs and also used in the food and textile industry and becomes an important avenue for the diversification of floriculture, sources of income generation and means of employment.

Besides, there is much scope for exploring diversified opportunities from different flowers. There is also enough scope to use the diversified-cum-value-added products for business purposes. These are bouquet preparation, bonsai, dried flowers and foliage, greeting cards, filler crops, wood rose, pine cones, draft wood, concept marketing, flower arrangements, flower garland, string, etc.

19.2.3 Essential Requirements and Available Infrastructural Facilities

Experience shows that investment decisions to enter floriculture may not be based solely on the love of flowers. It is a complex agri-business requiring a great deal of highly specialized knowledge and skills. The industry is highly technical and scientific. It is labour intensive and good management skills are essential which will enable to make a profit. A considerable time and study before making a business decision on the type of crop and location is necessary. The consumption basket is getting diversified towards value-added floral products and other by-products from flowers. Therefore, marketers must respond to these shifts in consumption. A business of value-added products will be successful when the floricultural producers carefully identify goods that utilize local resources and fill gaps in the market. Besides, as value addition increases cost of production, appropriate managerial strategies will increase the net cash return of small-scale floriculture enterprises.

19.2.4 Present Status of Government Initiatives

With the declaration of floriculture as an ‘extreme focus area by the Ministry of Commerce and Industry, Government of India, the floriculture sector has acquired a special status in the flower basket of India and is designated globally as a ‘Flower Power’. Six Agri Export Zones on floriculture have been set up in Sikkim, Tamil Nadu, Uttarakhand, Karnataka and Maharashtra. The APEDA has also taken a number of measures to facilitate floriculture exports. Some key Indian airports like New Delhi, Mumbai, Hyderabad, Bangalore, Chennai, Thriruvananthapuram and Cochin now have cold storage and cargo handling facilities. More airports will have these facilities in the future. Among other things, flower auction centres are also coming up in Bangalore, Mumbai, Noida, and Kolkata. These are readymade market facilities for trading and price discovery for a variety of flowers, both for export and domestic markets. India has to achieve the ambitious export target of ₹ 1000 crore per annum over the next 5 years, a paradigm shift is required (The key issues that need to be addressed in the Indian context are (1) economics of scale, (2) product range/latest varieties, (3) year-round exports, (4) quality control and (5) certification, cold chain management. The APEDA has been addressing these issues through various forums on a concerted basis given its mandate to promote floricultural exports from India (Anonymous 2010b).

19.2.5 Case of West Bengal

West Bengal is a unique state, where many commercial flowers are grown in different parts like the hills, plains, coasts, etc. In recent years, the government's department has advised farmers to examine the ways to take up the making of flower-based products, as the demand for basic flowers was falling. These products were expected to help farmers avoid losses or go in for distress sale of surplus flowers at rock-bottom prices. Some farmers were simply throwing away the crop in despair. The problem was that demand was flat as weddings were banned under the Hindu calendar during this time of the year and there were no festivals to boost demand. As a result, flower prices came down drastically. 'Bel' flowers were selling at ₹ 40–50 per kg in May and were now selling at ₹ 25 per kg. The 'jui' flower was going at ₹ 80 per kg in May and ₹ 30 per kg in June. 'Gladiolus' worth ₹ 20 per kg last month was now at ₹ 10 per kg. 'Rajanigandha' fetched ₹ 50–60 per kg in the wedding season and was now at ₹ 7–8 per kg (www.business-standard.com). So, value-added products have immense importance for commercial floriculturists. However, the State government has taken up several initiatives for development of the sector.

The State has set up a floriculture park at Mungpoo in North Bengal, which will provide a common infrastructure to the entrepreneurs. It will have cold storage facilities, grading/packaging rooms, tissue culture labs and refrigerated vans to collect flowers and then transport it. A mega flower mart in Mullickghat, Kolkata, is being set up at a cost of ₹ 250 million with dedicated flower stalls, state-of-the-art cold chain infrastructure for preserving flowers and plush stalls offering handicrafts and other ethnic treasures. There is comprehensive multi-storied flower market with facilities for cold storage at Panskura build at cost of more than US $ 1 million. The government has recently received a US $ 33 million proposal for setting up an open-air floriculture park on 200 acres mainly to cultivate roses at Rajarhat, Kolkata (Anonymous 2010c).

19.2.6 Value-Added Floriculture and Challenge

Of late, this sector is facing some challenges in the event of requirement of globalization and trade liberalization such as (1) growing of specific variety as per market preference, (2) site-specific appropriate technology for value-added products, (3) technical knowledge for standardization (Sanitary and Phyto-Sanitary Measures (SPS) under WTO), (4) synergy in the plan and approach of agencies at various levels of functional areas, namely research, extension, finance, quality assurance and certification (www.floriculturetoday.in).

19.2.7 Major Constraints

Developing countries have some inherent problems in respect of fulfilling the requirements of infrastructural facilities for producing market-oriented products. It is applicable for value-added floriculture also in India which are mainly related to trade environment, infrastructure and marketing. References suggest that there are many constraints which stand in the way of floriculture industry in the country. Out of that, the major constraints are (1) availability of basic inputs, including quality seeds and planting materials, quality irrigation and skilled manpower and aging plantations, (2) low availability of dedicated perishable carriers, higher freight rates, inadequate support infrastructure, constraint in achieving economies of scale and inadequate cold chain management, (3) product diversification and differentiation, vertical integration and innovation, (4) fragmented and vertically unorganized industry, (5) lack of post-harvest infrastructure and proper logistics support including cold chain mechanism, (6) inadequate infrastructure facilities for domestic market as well as at the export gateways, (7) financial institutions reluctant to support due to perceived risks and inadequate knowledge and (8) market intelligence unavailable on regular basis.

19.3 Suggestions

Based on the above discussion, the following suggestions may be put forward for further improvement of this sector. Appropriate steps may be taken to (1) design marketing initiatives to meet the market requirements, and yet remain competitive; (2) develop a state-of-the art integrated cold chain, for flowers right from the point-of-origin (growers) to the point of- consumption (customers) consistently (Hans 2011); (3) use advanced scientific technologies to maintain international quality standards; (4) develop indigenous technology for modern floriculture to suit the Indian growing conditions; (5) address the shortage of trained manpower on modern commercial floriculture, training centres and programmes for skill development in floriculture; (6) diversify product mix in the changing pattern of demand (domestic and export); (7) encourage the corporates and foreign direct investment (Ruud 2010).

References

APEDA (2011) Agricultural and Processed Food Products Export Development Authority (APEDA). www.apeda.gov.in

Anonymous (2000) Report of National Horticultural Board (NHB), Government of India, New Delhi

Anonymous (2009) Report of Export-Import Bank of India, No. 51, November. www.eximbankindia.gov.in . Accessed 22 Sept. 2010

Anonymous (2010a) Horticulture Today, Media Today Group, India, February Issue

Anonymous (2010b) Report of National Horticultural Board (NHB), Government of India, New Delhi

Anonymous (2010c) Report of Department of Horticulture and Food Processing, Government of West Bengal, Kolkata

Hans P (2011) The position of the marginal and small farmers in the value chain of cut flower in alluvial West Bengal, India and scope for an upgrading policy obtained from website prosenjit%20hans_1583.pdf

Ruud LM (2010) Floriculture worldwide: production, trade and consumption patterns show market opportunities and challenges, Wageningen University and Research Centre, Agricultural Economics Institute, The Hague, The Netherlands

References

[illegible]

Chapter 20
Role of Muga Culture in Diversification Strategy

K. K. Das, T. N. Roy and B. Das

20.1 Introduction

Silk is a natural fibre and regarded as one of the important items of aristocracy since pre-historic period. Even today, despite the onslaught of synthetic fibres, silk continues to reign supreme as the 'queen of textiles'. It is said 'Silk does for the body what diamonds do for the hand'. It is an item of luxury. India, China, Brazil, Thailand, Vietnam, Japan, Korea, etc. are the leading countries associated with production or trading of silk and silk goods. India, ranked second (Table 20.1) in raw silk production, earns a great amount of foreign currency each year through the export of various silk goods (₹ 3178 crores during 2009–2010 as per *Foreign Trade Statistics of India: Principal Commodities & Countries*, Kolkata, Anonymous 2010). It is the only country in the world which can produce all the five kinds of natural silk namely, mulberry, tasar, oak tasar, eri and muga. Of course, the bulk (about 80 %) of the total production is supplied from the fully domesticated silk, that is, mulberry silk (Anonymous 2007).

During the period from 2001–02 to 2009–10, *vanya* silk production grew at the rate of 15.40 per cent per annum (compared to 2.60 per cent in mulberry silk during same period) and therefore, share of non-mulberry silk in total raw silk production in the country is increasing day by day (Table 20.1). Muga silk, one of the three non-mulberry silks, produced by the silkworm *Antheraea assamensis,* is unique for its

K. K. Das (✉) · T. N. Roy
Department of Agricultural Economics, Uttar Banga Krishi Viswavidyalaya, Pundibari, Cooch Behar, West Bengal, India
e-mail: kkdas_ubkv@yahoo.com

T. N. Roy
e-mail: tnr_ubkv@yahoo.co.in

B. Das
Agricultural Sector Officer, Kadamtala, Agril. Sub-Division, North Tripura, Govt. of Tripura-799261, India
e-mail: banani5@yahoo.co.in

M. Ghosh et al. (eds.), *Diversification of Agriculture in Eastern India,*
India Studies in Business and Economics, DOI 10.1007/978-81-322-1997-2_20

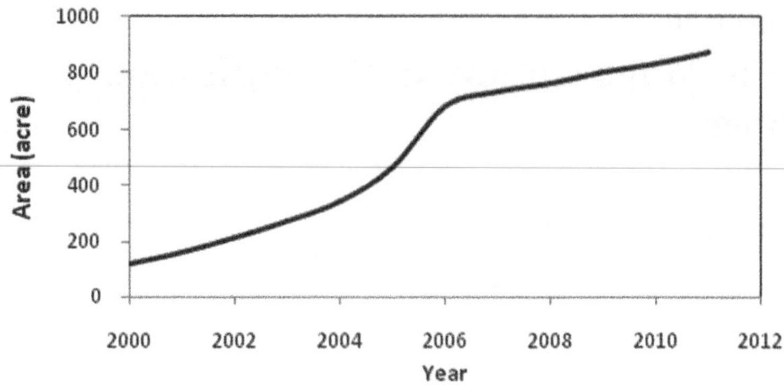

Fig. 20.1 Muga acreage in Cooch Behar

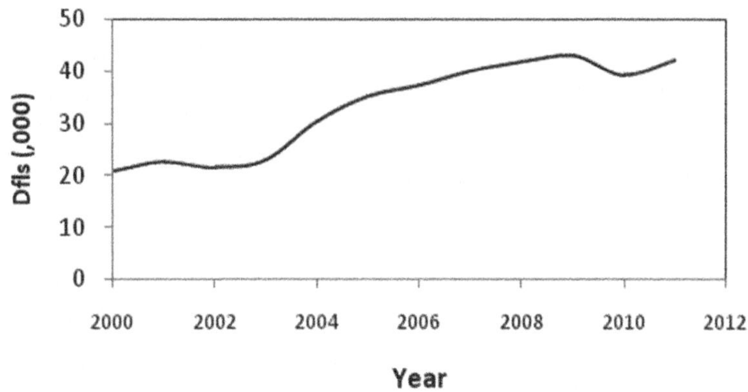

Fig. 20.2 Rearing of muga disease-free layings (dfls) in Cooch Behar

golden yellow colour. India is unique in producing this semi-domesticated silkworm. The Brahmaputra Valley area of Assam is famous all over the world for this yellow silk and shares about 94 per cent of total muga silk in India. It contributed about 3.06 % of raw non-mulberry silk production in the country in 2010–11 (Table 20.3)

Muga silk is produced on the fringes of the Brahmaputra Valley and is mainly confined to Assam. Cooch Behar, famous for its royal historical background, is a northern district of West Bengal, which lies on the fringe of the Brahmaputra Valley. It contributes about 0.2 per cent of muga silk production in the country (Table 20.4). Therefore, sensing the potentiality of climatic suitability of muga silkworm rearing in this district, attempts have been made to introduce muga culture in the district recently under the joint patronization of Directorate of Sericulture, Government of West Bengal, Central Silk Board, Ministry of Textiles, Govt. of India and NGOs. At present (2011–12), there are about 1800 households involved in muga culture with

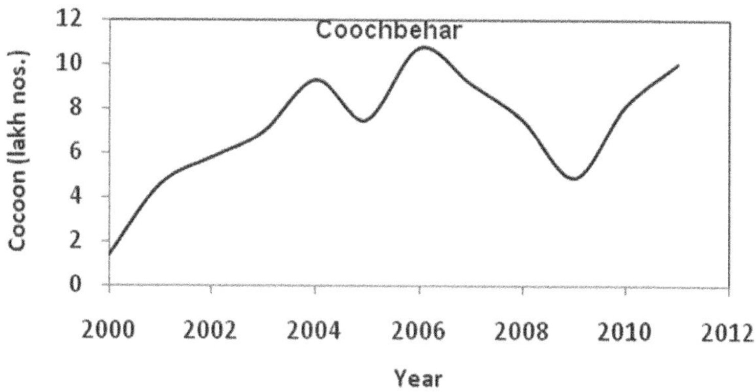

Fig. 20.3 Production of muga cocoon in Cooch Behar

Table 20.1 World raw silk production

Country	2007		2008		2009	
	Production (MT)	Percentage share	Production (MT)	Percentage share	Production (MT)	Percentage share
China	108,420	82.70	82.70	81.20	98,620	81.95
India	18,320	14.00	14.00	15.10	18,370	15.50
Brazil	1220	0.90	0.90	1.00	1177	0.60
Thailand	760	0.60	0.60	0.90	1100	0.50
Uzbekistan	950	0.70	0.70	0.70	865	0.60
Others	1505	1.10	1.10	1.10	1260	0.85

Source: International Sericulture Commission, D & B India

Table 20.2 Mulberry and non-mulberry silk production in India. (Source: Annual Reports of Central Silk Board, Ministry of Textiles, Government of India)

Year	Raw silk production (in tonnes)	
	Mulberry	Non-mulberry
1980–1981	4593 (91.11)	448 (8.89)
1985–1986	7029 (89.01)	868 (10.99)
1990–1991	11,486 (91.45)	1074 (8.55)
1995–1996	12,884 (92.63)	1025 (7.37)
2000–2001	14,432 (91.01)	1425 (8.99)
2005–2006	15,445 (89.25)	1860 (10.75)
2010–2011	16,360 (80.16)	4050 (19.84)

Figures within parentheses indicate percentage of total raw silk production

an extent of about 874 acres of muga host plantation (Figs. 20.1, 20.2, and 20.3). Now, the question comes, will the venture be a remunerative one for the farming folk of this area or not? Therefore, an attempt has been made in this study to know about the 'economics' of Muga culture at the farmers' level in this district.

Table 20.3 Non-mulberry silk production in India

Year	Raw silk production (in tonnes)		
	Tasar	Eri	Muga
1980–1981	265 (59.15)	135 (30.13)	48 (10.72)
1985–1986	464 (53.46)	352 (40.55)	52 (5.99)
1990–1991	380 (35.38)	624 (58.10)	70 (6.52)
1995–1996	194 (18.93)	745 (72.68)	86 (8.39)
2000–2001	237 (16.63)	1089 (76.42)	99 (6.95)
2005–2006	308 (16.56)	1442 (77.53)	110 (5.91)
2010–2011	1166 (28.79)	2760 (68.14)	124 (3.06)

Source: Annual Reports of Central Silk Board, Ministry of Textile, Govt. of India
Figures within parentheses indicate percentage to total non-mulberry raw silk production

Table 20.4 State-wise muga silk production in India. (Source: Annual Reports of Central Silk Board, Ministry of Textile, Govt. of India)

State	Muga silk production (in tonnes)		Percentage change
	2009–2010	2010–2011	
Assam	93.00 (88.57)	117.00 (94.35)	(+) 25.81
West Bengal	0.20 (0.19)	0.25 (0.20)	(+) 25.00
Other North Eastern States	11.80 (11.24)	6.75 (5.46)	(–) 42.80
India	105.00	124.00	(+) 18.10

Figures within parentheses indicate share of corresponding state in muga silk production

20.2 Data and Methodology

Muga culture has spread to about ten blocks of Cooch Behar district, of which Dinhata-I and Coochbehar-II blocks hold the bulk of the rearers. Therefore, the present study has been taken up basing on micro-level information collected through door-to-door surveys from these two sample blocks of Cooch Behar district. A total of 50 rearers (25 from each block) have been covered for the present study. The villages in each block are selected on the basis of availability/intensity of muga households there and finally, the muga rearers are selected randomly. For the present study, data pertaining to one agriculture year, that is 2008–2009 are obtained by the 'survey' method. As because there are the practices of undertaking rearing of muga silkworm during four seasons, namely *Chotua* (Seed, P1), *Jethua* (second commercial), *Bhodia* (Seed, P1), *Kotia* (chief commercial). Collection of data has been made in four rounds, once in the end of every rearing season. The study aims to judge the suitability of the avocation from various 'economic' aspects. Therefore, in order to work out the 'economics' of Muga silkworm rearing, the cost of production of Muga cocoon as well as of leaf of host plant (i.e. som; *Machilus bombycina* King) have been worked out by using the feasible standard cost concepts. Simple statistical methodologies have been employed to work out the 'economics'.

Table 20.5 Important traits of the muga rearers

Traits/features	Unit of expression	Dinhata-I	Cooch Behar-II	Average of two blocks
Family size	Number	5.15	4.67	4.86
Education[a]	Score	1.51	1.31	1.43
Male: female	Ratio	1000:544	1000:926	1000:667
Age of family head	Years	54.30	46.40	50.41
Farmers practising only muga culture	Percentage	45.0	46.67	45.71
Farmers practising muga culture with farming	Percentage	50.0	53.33	51.43
Size of som garden	Acre	0.80	0.67	0.77
Percentage of area under som garden	Percentage	37.04	42.00	39.14
Extension contact[b]	Score	2.85	3.06	2.94
Distance of som garden from the homestead area	Metres	63.75	249.00	143.00
Number of som plants	Numbers/acre	711	651	684

[a] Data about educational background have been elicited by obtaining educational score of a farm family through *weighted average method.* While calculating, score has been assigned to each of the farm family member in the following manner: illiterate=0; up to Class IV=1; Class V–Class X=2; above Class-X=3

[b] To extract the answer to this probing question about extension contact, the muga rearers have been asked '*whether they have visited sericulture offices, any bank or financial institutions, other offices or participated in any "Krishi Mela", workshop, seminars, etc.*'. The corresponding institute or programme is assigned the score 1 if the farmer visited or attended it, 0 otherwise

20.3 Results and Discussion

As muga culture is of recent origin in the area, its progress (both horizontal and vertical) is supposed to be influenced by the socioeconomic traits like education, mobility, income and expenditure, etc. A brief idea on important social and economic parameters can be had from the perusal of Table 20.5 which depicts that the Muga rearers are at a relatively disadvantageous position, especially on the question of perception and adoption of improved technologies on maintenance of som/soalu (*Litsea polyantha* Juss.) plantation (the host plant for Muga silkworm) and also on Muga silkworm rearing. They not only have relatively poor educational backgrounds (up to Class VI/VII, on an average) but also have fewer female counterparts (two females per three males) to take part in rearing activities, which is an absolute essential (Saraswathi and Sumangala 2001) for making the venture economic. The average age of the farm family head (who have poor education only up to Class III/IV) is more than 50 years, which is not at all favourable for improving adoption capability; the score on extension contact is below 3.0 too. About 50 % of the muga farm families do take up muga rearing along with the farming of other crops and allocate 39 % of total land holding for raising the muga silkworm's host plant (som/soalu plantation). Thus, the average size of a som/soalu garden becomes 0.80 acre, and whereas the gardens in Dinahata-I are adjacent (65–70 m) to the residence,

Table 20.6 Seasons of muga silkworm rearing in Cooch Behar district, West Bengal

Crop	Season	Month covered	Status	Date of brushing
Jarua	Winter	Dec–Feb	Pre-seed (P-2)	17–20th Jan
Chotua	Early spring	Mar–April	Seed (P-1)	25–28th Mar
Jethua	Spring	May–June	Second commercial	14–17th May
Aherua	Early summer	July	Pre-seed (P-2)	3–6th July
Bhodia	Late summer	Aug–Sept	Seed (P-1)	22–26th Aug
Kotia	Autumn	Oct–Nov	Main commercial	11–15th Oct

Sources: Chakravorty et al. 2008; Das and Das 2008

those of Coochbehar-II remain far away (around 250 m) from it. This length of distance may play a critical role in 'economic' application of inputs and the managerial ability of the muga rearers. These two attributes are supposed to have inverse association with distance of farm field from the farm household. The size of a muga farm family is around five members.

Muga silkworm is a mono race. Though the primary food plants of muga silkworms are Som (*Machilus bombycina* King) and Soalu (*Litsea polyantha* Juss.), rearing is undertaken mainly on the leaves of som plants in this area. The muga silkworm is multivoltine in nature and 5–6 crops can be raised in a year in general. The crop cycle (i.e. rearing seasons) of this commercially exploited semi-domesticated silkworm in Cooch Behar is presented in Table 20.6. Out of these six crops (i.e. rearing seasons), four crops, namely, *Chotua, Jethua, Bhodia* and *Kotia* are undertaken by the muga rearers of Coochbehar district; while the seasons *Jethua* and *Kotia* are undertaken as commercial crops, *Chotua* and *Bhodia* are undertaken as seed crop (Table 20.6). The seed crops like *Jarua* and *Aherua* are hazardous due to various biotic and abiotic factors. These two crops are not undertaken by the rearers of this area.

Input utilizations in the muga garden do vary from season to season and therefore, the pattern of input utilization in muga garden is studied season wise. Muga rearers, in this locality, raise and maintain som plants to supplement the silkworms with food (i.e. leaves). Actually, the muga food plants were never utilized to the maximum level and one full-grown tree of 12–20 years' growth can support the rearing of 5–10 layings and yield 500 cocoons in one season and one tree can be utilized for two rearings in a year alternatively during spring and autumn (Handbook of Muga Culture 1988). On an average 684 som plants are raised in a garden of 1.0 acre. Comparing to the standard spacing, the plants are bit densely maintained here. But the whole plantation is utilized alternatively in the feasible rearing seasons. The customary practice is to utilize the som leaves from a certain part of the garden in one season keeping the remaining part unutilized: about 56–58 % of the existing som plants in a muga garden are utilized during the commercial seasons (i.e. *Jethua* and *Kotia*) while 44–48 % of the plants are utilized in seed crops, that is, *Chotua* and *Bhodia*. Management capabilities and resource constraints are thought to be behind this differential nature in utilization of existing som plants in the garden.

Inorganic fertilizers namely, nitrogen, phosphorous and potash (NPK) are applied in the som garden to maintain the health and vigour of the plants. Urea is the main source of nitrogen. Single super phosphate (SSP) is the main source of phosphorous while muriate of potash (MOP) happens to be the main source of potassium in this area. These three nutrients are applied once in every season. Generally, fertilizer is applied during land preparation operation in and around the som plants. Fertilizer is applied through deep placement method by making three to four holes around the tree at least 1 foot away from the base of the muga plant. On an average, fertilizers (Urea, SSP and MOP) are applied in the ratio of 3:4:1 (143:183:43 g) per plant (Table 20.7) which is far below the recommended ratio of 550:750:200 g (Thangavelu et al. 1988). One of the serious drawbacks in muga culture is that the plants are not cared for properly. The existing case does suggest from the very fact that fertilizers are used in abysmally low quantity. Thus, there may be fewer amounts of leaves to supply the food material of muga silkworms. As such, no seasonal variation is observed in fertilizer application.

Of the four rearing seasons, irrigation water is supplemented in the som garden only during Chotua season, in general (Table 20.7). The plants are affected by insect/pests such as stem borer, sucking pests, leaf miner, leaf gall, etc. Rearers, in general, do use protection chemicals for protection of food plant. Also disinfection is done during *Chotua* season. About 100 ml of a chemical in the trade name of '*oostad*' is used once in every season to spray in the som garden for protection against disease and pests. Bleaching powder is used as disinfectant (Table 20.7).

Like other lepidopterans, muga silkworm is a holometabolous insect, passing through a complete metamorphosis from egg (*koni*) to adult (*chakari*) stage through two intermediate stages of larva (*polu*) and pupa (*leta*). The entire life cycle lasts for about 50 days in summer and 120 days in winter. Rearing of the muga silkworm, which is semi-domesticated in nature, involves a number of operations—procurement of disease-free layings (dfls), brushing of larvae, maintenance of larvae, transferring of larvae, collection of mature worms for cocooning and harvesting of ripe cocoons, to name a few. As stated earlier, rearing of muga silkworm is undertaken mainly in four seasons of which *Kotia* and *Jethua* are treated as commercial seasons, the other two as seed crops. Accordingly, a greater number of dfls is reared in these two commercial seasons. The rearers of this area reared about 507 dfls in *Jethua* and 612 dfls during *Kotia* to feed 1.0 acre of som garden. About 447 dfls are reared in *Bhodia* seed crop and 381 dfls are reared in *Chotua* seed crop (Table 20.8). The dfls, which are procured from private grainures during commercial crops and from government sources during seed crops, are very much costly. During commercial seasons, procurement price of 100 dfls of muga silkworm remains at ₹ 450 whereas it goes up to ₹ 500 during seed crops.

Human labour forms a major component of both the maintenance of som plantations as well as in the rearing of silkworms. An idea about the utilization of human resources in the som garden can be had from Table 20.9. Human labour in som garden is used for inter-cultivation of land, weeding, training, pruning operation, etc. In these operations, male dominance is mostly seen. Both labourers from within the farm family as well as hired (on daily basis) labourers are employed by the muga rearers. It is observed that on an average 10–16 man-days are employed seasonally for inter-cultivation of an acre of muga garden (Table 20.9). Of these, about 45 % is

Table 20.7 Input utilization for the maintainance of muga host plant

Items	Unit of application	Application of inputs				
		Jethua	*Bhodia*	*Kotia*	*Chotua*	Annual/average
Number of plants utilized	Numbers/acre	662	620	677	640	650
Quantity of organic manure	q/acre	–	28.02 (4.06)	–	27.33 (3.98)	55.35 (8.04)
Nitrogenous fertilizer	kg/acre	24.63 (35.86)	24.60 (35.86)	24.00 (35.86)	24.45 (35.86)	97.68 (143.44)
Phosphatic fertilizer	kg/acre	31.47 (46.14)	33.00 (46.14)	31.74 (46.3)	29.46 (44.57)	125.67 (183.15)
Potassic fertilizer	kg/acre	8.10 (11.29)	7.50 (10.40)	7.74 (10.57)	6.81 (10.57)	30.15 (42.83)
Irrigation hours	h/acre	–	–	–	10.99	10.99
Disinfectants for host plants	kg/acre	–	–	–	21.57 (41.97)	21.57 (41.97)
Plant protection chemicals	mL/acre	231.27	152.58	234.00	326.91	944.76
Human labour	Man-days/acre	29.07	36.78	31.95	40.74	138.54

Figures within bracket indicate input application per plant

Table 20.8 Input utilization for the rearing of muga silkworm

Items	Unit of application	Application of inputs				
		Jethua	*Bhodia*	*Kotia*	*Chotua*	Annual/average
Muga disease-free layings (dfls) reared	Number/acre	507	447	612	381	1947
Human labour for placement of muga silkworms in the host plant and other related operations	Man-days/acre	5.31	3.36	4.35	3.45	16.47
Human labour for harvesting of cocoons	Man-days/acre	4.11	4.29	5.13	3.84	17.37

supplied from within the family. Annually, about 45 mandays are employed for this purpose and employment of mandays is more in the commercial seasons. Almost similar is the case (9–10 man-days) for weeding operation in the garden.

Muga food plants need proper training and pruning. It is done with the objective of increasing the foliage, removing the diseased parts, to give desirable shape to the plant, to control the height of the plant, etc. Without training, the leaves of muga food plant become rough, coarse and small and damaged by diseases and insect pests. Plants become tall and inconvenient for management of silkworms rearing. Generally, the training is done once in a year, preferably before the onset of monsoon rain (i.e., before the *Bhodia* rearing season). It can be observed from Table 20.8 that about 38 man-days are employed annually for both these operation. However, the peculiarity in this case is that, the majority (about 65 %) of the total labour is supplied from within the family.

The tradition-bound muga rearers conduct muga culture in a very pious and religious manner. All people in the house participate in muga silkworm rearing, whereas muga silk reeling and weaving are mostly done by women. Rearing of muga silkworms involves brushing of larvae, placing of worms in the som/soalu plants and continuous vigil on the tiny worms, handling of moulting larvae, transferring of larvae, preparation for cocoonage, collection of mature worms and placing them in the cocoonage (*jali*), harvesting of cocoons, etc. The jobs are hazardous and deserve careful nurturing. Both female and male labourers are used in these operations. An idea about the man-days used can be observed from Table 20.9. It is found from this table that about 16 man-days are utilized annually to place the hatched tiny larvae on muga plants and keeping vigil on them. Obviously, the required man-days are more in the commercial seasons than the seed crops. Similarly, about 17 man-days are utilized for collection of cocoons. Thus, in total the avocation needs employment of 172 man-days for rearing of muga silkworms from an acre of som/soalu garden.

Some plants are perennial in nature and its leaves are utilized differentially in different rearing seasons. It is estimated that a total sum of about ₹ 15,134 is required for establishment of 1 acre of som garden (Department of Sericulture, Government of West Bengal, Coochbehar). An additional estimated annual sum of ₹ 20,189.81 is required for maintenance of that som garden and the subsequent rearing of muga

Table 20.9 Operation-wise mandays application in rearing of muga silkworms

Major item	Items	Application of inputs				
		Jethua	*Bhodia*	*Kotia*	*Chotua*	Annual/average
For maintenance of muga host plants	Intercultivation	10.89	15.51	10.95	7.50	44.85
	Weeding and cleaning	10.14	9.15	8.61	9.48	37.38
	Application of manures and fertilizer	1.50	4.83	1.50	6.33	14.16
	Irrigation	–	–	–	3.78	3.78
	Training and pruning	6.54	7.29	10.89	13.65	38.37
	Sub-total	29.07	36.78	31.95	40.74	138.54
For rearing of muga silkworm	Placement of muga silkworm in host plants	5.31	3.36	4.35	3.45	16.47
	Collection of cocoon	4.11	4.29	5.13	3.84	17.37
	Sub-total	9.42	7.65	9.48	7.29	33.84
	Total	38.49	44.43	41.43	48.03	172.38

Table 20.10 Season-wise cost and return prospect in muga culture

Major head	Sub-head	Season-wise cost and return (₹/acre)				
		Jethua	*Bhodia*	*Kotia*	*Chotua*	Annual/average
Cost	Prime cost	3489.39	5412.14	4108.15	7180.13	20,189.81
	Cost A1	3954.30	6062.43	4609.22	7909.10	22,535.05
	Cost C	4659.30	6951.99	5442.99	9076.07	26,130.35
Return	Gross return	8106.42	10,297.30	11,095.51	11,575.79	41,075.02
	Return over prime cost	4617.03	4885.16	6987.36	4395.49	20,885.21
	Return over Cost A1	4152.12	4234.87	6486.29	3666.69	17,003.57
	Return over Cost C	3447.12	3345.31	5652.52	2499.72	14,944.67
Return per Rupee investment		2.32	1.90	2.70	1.61	2.03

silkworm (Table 20.10). Three types of maintenance cost namely, Prime Cost, Cost A1 and Cost C have been calculated. The monetary cost of rearing muga silkworm inclusive of maintenance cost of muga host plant garden constitutes the Prime Cost. The idea about Cost A1 has been arrived by adding interest on working capital at 12 %, depreciation and repair of farm tools and machinery, apportioned part of Establishment Cost for setting up muga garden with Prime Cost. The value of Cost C is obtained by adding imputed value of family labour with Cost A1.

Of course, there remains seasonal variation in maintenance cost of the garden as the number of dfls reared varies according to seasons. More dfls are reared in commercial seasons that are in *Jethua* and *Kotia*. These two commercial seasons share about two thirds (65 %) of the total number of dfls reared in a year. On an average, the muga farmers of this district rear a total of 1947 dfls per acre in a year and harvest, on an average, 42 cocoons per dfl.The total number of cocoons per muga dfl in seed crop is slightly more than that of commercial crops. This difference in cocoon production is supposed to have great impact on the question of 'economization'. On average a muga rearer earns an annual gross sum of ₹ 41,075. Margin of net return has been calculated from important consideration— over Explicit/Prime (Material) Cost, over Cost A1 and Cost C. A rearer of Cooch Behar district with a 1-acre som plantation can earn an annual income of ₹ 20,885.21 over Prime Cost, ₹ 17,003.57 over Cost A1 and ₹ 14,944.67 over Cost C (Table 20.10).

Rearing of muga silkworms, which are wild in nature, requires constant supervision, close vigil and careful nourishment since the very first day of larval stage of this tiny worm. Situation of som gardens within very little distance from home (50–65 m) may make it convenient to undertake these jobs with ease. This also makes it easier for the family members (especially, women) to attend the rearing as and when required. The muga rearers with relatively higher family size and more male members are found to support the labour-intensive muga culture more efficiently. Finally, we turn towards the prospect of return from the rearing of Muga silkworm avocation per rupee investment. A perusal of Table 20.9 shows that maximum return prospect is available in *Kotia* which is main commercial crop season in this area. On average, a rearer can earn ₹ 2.70 from a ₹ 1 investment in this season. The rearers of Dinhata-I block receive nearly three times of their rupee investment in this season which is considered to be high in any standard. Return prospects in different season are also displayed in the table.

20.4 Conclusion and Policy Implication

It can be said that muga silkworm rearing is a good farming avocation which not only fetches very good return but also can generate quite a large human labour employment and proves it a good 'economic' prospect. Therefore, it has capacity to afford better livelihood opportunity for the farming folk of this area and has the potential to spread further in future. But, the avocation suffers from the following constraints which need due consideration for its progress and development: (1) lack of Pure Line to identify high-yielding ones; (2) low fecundity coupled with low to very low hatching during *Aherua* and *Bhodia* crops; (3) weakly integrated seed multiplication chain at P2 and P1 levels; (4) frequent crop loss due to adverse climatic conditions affecting seed multiplication; (5) inadequate local cold storage facilities for skipping rearing during adverse seasons.

References

Anonymous (2007) Annual Report. Central Silk Board, Ministry of Textiles, Government of India, Bangalore

Anonymous (2010) Foreign trade statistics of India. Principal Commodities & Countries, Government of India, Kolkata

Chakravorty R, Dutta P (2008) Problem of muga culture Antheraea assamensis in Coochbehar and intergrated management approach. Paper presented at Workshop on Disease and pest management organized by Central Muga and Eri Training Institute, Jorhart on 29 Feb 2008 at Cooch Behar, West Bengal, pp 1–14

Das PK, Das SK (2008) Status and strategies of muga silkworm seed production in West Bengal. Paper presented at Workshop on Disease and pest management organized by Central Muga and Eri Training Institute, Jorhart on 29th Feb, 2008 at Coochbehar, West Bengal, pp 73–77. http://www.coochbehar.nic.in/HTMfiles/CoB_Seri_Muga.html#muga_statistics

Saraswathi JM, Sumangala PR (2001) Participation of farm women in sericulture enterprise. Indian J Sericult 40(1):86–91

Thangavelu K, Chakraborty AK, Bhagowati AK, Isa M (1988) Handbook of muga culture. Central Silk Board, Bangalore

Index

M. Ghosh et al. (eds.), *Diversification of Agriculture in Eastern India*,
India Studies in Business and Economics, DOI 10.1007/978-81-322-1997-2

The manufacturer's authorised representative in the EU is Springer Nature Customer Service Centre GmbH, Europaplatz 3, 69115 Heidelberg, Germany. If you have any concerns regarding our products, please contact ProductSafety@springernature.com

Printed and bound by CPI Group (UK) Ltd, Croydon, CR0 4YY
28/09/2026
02224088-0012